POSTMODERN WAR

CRITICAL PERSPECTIVES
A Guilford Series
DOUGLAS KELLNER, Editor
University of Texas at Austin

A THEORY OF HUMAN NEED
Len Doyal and Ian Gough

POSTMODERN THEORY: CRITICAL INTERROGATIONS
Steven Best and Douglas Kellner

PSYCHOANALYTIC POLITICS, SECOND EDITION: JACQUES LACAN
AND FREUD'S FRENCH REVOLUTION
Sherry Turkle

POSTNATIONAL IDENTITY: CRITICAL THEORY AND EXISTENTIAL PHILOSOPHY
IN HABERMAS, KIERKEGAARD, AND HAVEL
Martin J. Matuštík

THEORY AS RESISTANCE: POLITICS AND CULTURE AFTER (POST)STRUCTURALISM
Mas'ud Zavarzadeh and Donald Morton

POSTMODERNISM AND SOCIAL INQUIRY
David R. Dickens and Andrea Fontana, Editors

MARXISM IN THE POSTMODERN AGE: CONFRONTING THE NEW WORLD ORDER
Antonio Callari, Stephen Cullenberg, and Carole Biewener, Editors

AFTER MARXISM
Ronald Aronson

THE POLITICS OF HISTORICAL VISION: MARX, FOUCAULT, HABERMAS
Steven Best

ROADS TO DOMINION: RIGHT-WING MOVEMENTS AND POLITICAL POWER
IN THE UNITED STATES
Sara Diamond

LEWIS MUMFORD AND THE ECOLOGICAL REGION: THE POLITICS OF PLANNING
Mark Luccarelli

SIGN WARS: THE CLUTTERED LANDSCAPE OF ADVERTISING
Robert Goldman and Stephen Papson

REVOLUTION OF CONSCIENCE: MARTIN LUTHER KING, JR.,
AND THE PHILOSOPHY OF NONVIOLENCE
Greg Moses

POSTMODERN WAR
Chris Hables Gray

POSTMODERN WAR
THE NEW POLITICS OF CONFLICT

Chris Hables Gray

THE GUILFORD PRESS
New York London

© 1997 The Guilford Press
A Division of Guilford Publications, Inc.
72 Spring Street, New York, NY 10012

Printed in the United States of America

This book is printed on acid-free paper.

Last digit is print number: 9 8 7 6 5 4 3 2 1

Library of Congress Cataloging-in-Publication Data

Gray, Chris Hables
    Postmodern war: the new politics of conflict / Chris Hables Gray.
        p.   cm.—(Critical perspectives)
    Includes bibliographical references and index.
    ISBN 1-57230-160-0 (hard).—ISBN 1-57230-176-7 (paper)
    1. World politics—1989–. 2. War—Forecasting. 3. Military
history, Modern—20th century. 4. War—Psychological aspects.
I. Title. II. Series: Critical perspectives (New York, N.Y.)
D860.G75   1997
306.6'6—dc21   97-1538
                                                                CIP

To Carl Harp, cofounder of Men Against Sexism,
Walla Walla Federal Penitentiary. Dead of state terrorism.
Love and rage, Carl.

Preface

While the arguments of this book may seem contentious, they actually rest on many agreements. I agree with Noam Chomsky that there is "a persistent fair probability of nuclear war" (1986, p. 39). I agree with Richard Lebow and Janice Stein that we all lost the Cold War (1994), although we did win the one thing that we needed most, time. I agree with many specific things that historians and others have said, and I cite them at length in this book. But there are a number of intellectual dogmas that I don't ascribe to.

I reject simplistic accounts of the Cold War, and of all wars. In the book that follows I challenge much received wisdom. If you disagree, I'd love to hear from you. Write me care of this publisher.

There is one dogma, of the history profession, that I want to challenge here in the Preface, as it is a meta-issue not really addressed in the main text. It has to do with the idea of progress. I believe in progress, of a sort, even though history has its cycles. History does repeat itself, as tragedy and farce, but not always. New things happen. Many historians argue against the idea that history can be progressive; that it is going somewhere. But I disagree with them. There are patterns we can see that clearly show a kind of progress, if we remember that love and cancer both progress, grow, and spread. Technoscience, and the human impact on nature, where we can see progress of some type clearly, is obviously a mixture of healthy growth and something else. War is part of that something else. This is an account of its progress.

It is contested as are all stories. There is a place for stories in academia (Cronon 1992) and you will see that I mainly think in stories. For me, there are stories in everything. For example, I have dedicated this book to Carl Harp. Now there's a story, several stories. There's a story about Carl Harp, the beautiful anarchist activist who formed a Men Against Sexism group in prison and who was eventually murdered by the forces of order, as he called them. I was fortunate enough to meet Carl a few times, and learned a great deal from him. And there is a story to the dedication itself.

Originally I dedicated this book to another wonderful person, a deci-paracida in South America, presumed long dead in Argentina's Dirty War. But then, years later as this book was in press, I discovered he had recently come out of hiding in exile and that he had survived. So, a happy ending, in part. Just as war has its happy moments, especially in the stories of those who seek to end it.

The nature of these forces who seek to wage peace is an important part of the story of our present. The incredible mobilizations of women in countries as different as South Africa and Bosnia is one crucial example, and another happy story in the larger sad one of war (Lederer, 1995). I discuss peace activism at length towards the end of this book because it is part of the war story, as much as the effectiveness of machine guns or the madness of mutually assured destruction. There are many stories.

Science, which has offered us so many difficult tests in the area of war, also offers us one of these happy stories, hope in the power of life. Whether or not Rock 84001 really has signs of ancient life from Mars, the hope it offers is real. We might find life on other planets. We might have already. We might make peace on our planet. Peace could be breaking out all over. There is always hope, where there is life. Life writes stories just by living. Humans have formalized this, of course. Hence this book. I hope it holds your attention.

LOVE AND RAGE,

CHRIS HABLES GRAY, 8/16/96

Acknowledgments

I would to first thank Donna Haraway, my mentor. She is an inspiration. I would also like to sincerely thank the other members of my dissertation committee: Barbara Epstein, Terry Winograd, and Bruce Larkin, for their support and advice. Billie Harris, the administrator of the History of Consciousness Board during my time there, was an invaluable help, a steadfast supporter, and a good friend. Everyone at HistCon deserves thanks. It is a unique, wonder-filled program and a weird but real community. But I would especially like to mention Hayden White, Jim Clifford, and my good friends Zoe Sofia, Paul Edwards, Sharon Helsel, Noel Sturgeon, Sarah Williams, and Ron Eglash. Everyone in our Cultural Studies of Science and Technology Research Group has been helpful, thoughtful, and amusing, especially Mark Driscoll.

I owe a great deal to the faculty and students I've worked with at the University of California, Santa Cruz, and in other places as distant as Moscow, but I don't have space to name them all here. I also must thank the Institute on Global Conflict and Cooperation of the University of California at San Diego, the Universitiy of California Regents, the Eisenhower Fellowships, and the Smithsonian Institution for funding support. Paul Ceruzzi of the Smithsonian was a kind and gracious host during my summer there. James Der Derian as a commentator and Les Levidow, an editor, helped me improve this book a great deal, as did Hugh Gusterson, a fine friend and a fine reader. Of course all errors are my responsibility.

Finally, I could not have done this work without my friends and family. Jane, Corey, Zachary, Mary Hables, the Wilsons, the Grays, and all my companeros, you know who you are, thank you!

CONTENTS

Introduction: From Sarajevo to Sarajevo, the USS *Oklahoma* to Oklahoma City

"You may not be interested in war, but war is interested in you." This quotation, ascribed by some to Leon Trotsky and others to Leo Tolstoy, is very popular among historians of war. I admit I've used it myself in my more irritable moments, not just in texts but with friends and students alike. Why is it so seductive a . . . story? Several reasons are clear. First, these are very confusing times when it comes to war. From domestic terrorism to nuclear war the range of contemporary war is boggling—and baffling. Second, war's restless energy, its almost palpable agency, are also clear in this little bon mot. War isn't just something that humans *make*, it *is*.

Or is it? Explaining what makes war possible is one of the two great goals of this book. Well, perhaps that is putting it optimistically. Perhaps telling several intertwined stories about how war works would be more realistic—and quite sufficient. The second main goal of this book is to specifically explain, in part, why war has become the confusion it is now and to shed some light on what we can expect in the future. I am convinced that there are ways to explain contemporary war. The best of these stories are grounded in careful attention to what we can know of history. The cloudy mirror of the future is a reflection and projection of the past. This explains my emphasis on history. Too many theorists of contemporary war are unfamiliar with the long twisted tale of war itself.

So this is the story of war today, in reference to what war has been and dreams of what it will be. The rest is just elaboration. The postmodern theory, the medieval history, the arcana of high-tech weaponry and even the exegesis

[1]

of official infowar doctrines are all just ways of mapping this tale—the most useful mappings for me, although I hope you too will find some helpful, dear reader. War is a living text, after all, and we are all of us bound into it, of it, even as we tell our parts, as it writes our future.

> First World, ha, ha, ha!
> —*Popular chant at the pro-Zapatista rally*
> *in Mexico City, March 1994*

> Here we are, the dead of all times, dead again, but now
> to live!
> —*Zapatista communiqué, January 1994*

One history of our century is the simple, sad story of 1914 to 1995—Sarajevo, Bosnia–Herzegovina. Through this lens all of modern Western history, a thousand years, could be perceived in the tortuous tale of this city with its incomplete colonializations, its shifting governments, its interminable ethnic, religious, imperial, and ideological wars. Certainly the "sameness" of it all is horrifying. This is true as well if one looks around the world, a few short years since the end of the all-defining capitalist–communist conflict. Then the many wars around the world were understood to be part of this great Manichaean struggle for the future of civilization: red versus red, white, blue. Now the Cold War is won, and yet wars proliferate all the same. And now, without the Cold War framework, these wars are revealed to be in many respects nightmares out of the past: Balkan and Transcaucasian wars of rape,[1] murder, ethnic cleansings, and occupation; tribal wars in Africa and Southwest Asia; brutal repressions and peasant rebellions in Central and South America; Amerindian uprisings in Ontario, the Amazon, Chiapas; nationalist terrorism in Europe and Asia; oil wars in the Middle East; and extremist (religious and ideological) terror in Japan gas attacks and U.S. fertilizer bombs; fundamentalism in state power in parts of Africa and Asia; religious strife everywhere.

Samuel Huntington has been so struck by this proliferation of ancient conflicts and hatreds that he now posits that the wars of the future will be caused by the "Clash of Civilizations," the model being, one supposes, the ancient world of Rome and perhaps the Crusades as well.[2]

But is it so simple? Despite the disconcerting continuity between many contemporary conflicts and ancient, brutal quarrels, any close observer will also note gaping discontinuities. There are some very new and unique aspects to war today that represent a tremendous break from war's history and also point to war's possible futures.

Let us look at some details.

- Within the next 20 years it is highly probable that there will be the use of nuclear weapons in a regional and/or terrorist conflict. The proliferation of nuclear technology and the tenacity of war culture make this a very safe prediction. Not only will nuclear weapons continue to proliferate, but ballistic missiles will as well (Nolan and Wheelon, 1990). It is also just as likely that there will be at least one major use of chemical or biological weapons. As small nations and nation-wanna-bes seek access to these "poor man" weapons of mass destruction, already being deployed domestically in Japan, the major powers continue to perfect them (Pringle, 1993; Wright, 1993). The underlying issue here, and the driving dynamic of contemporary war, is the proliferation of military technologies.
- The growing strain of continuing human expansion and industrialization on the environment is a major, and increasing, cause of conflict in the world. It is no accident that the Chiapas rebellion is in one of the last rain forests, the Selva Lacandona, in the world (Langelle, 1995). Resource wars have always been common and continue to be, from storming armored columns in the Gulf War to armed resistance to wolves in the national parks of the United States. This whole range of conflicts will increase as resources continue to be depleted while the requirements of the human population of the world increase. The same dynamic underlies the growing danger of a worldwide pandemic.
- The global increase in human communication has fundamentally changed both the nature of local conflicts as well as the potentialities for peace. International networking is increasing. The UN may appear sadly ineffectual in many cases, but it remains a remarkable attempt to control war. The internationalization of human culture is producing many important effects, from multinational companies (resistant as never before to the powers of nation-states) to the growth of what the Zapatistas call "international civil society." What it may lead to is not just unclear, it has yet to be determined.
- But that hasn't put a crimp into meta-analysis, such as Samuel Huntington's "Clash of Civilizations" described above and, to be honest, this very book. Two Rand prognosticators, for example, have gone so far as to proclaim that "Cyberwar is Coming!" and they have carefully redefined existing notions of "information war" into various classes of cyberwar and netwar (Arquilla and Ronfeldt, 1993). But for them, as for most war theorists, the goal is to improve war, not to understand war itself in the context of contemporary society. The broader implications, from horrific to idyllic, of war's latest evolution do not concern them.

But that is the concern of this book. War is not just in transition, it is in crisis. What it will become will determine not just some aspect of the human future but whether or not we have one. There are some reasons for optimism and many for terror, but the story itself starts with confusion,

confusion over the very idea of war. Put crudely, "Is it nuclear submarines? Fertilizer bombs? Genocide? Peacekeepers? All of the above?"

In April 1994 the USS *Oklahoma*, the most advanced nuclear hunter-C killer submarine of the U.S. Navy, was relaunched after her major multimillion dollar upgrade and overhaul (Wood, 1994, p. A3). Her mission: to destroy "boomers," enemy ballistic submarines, none of which threaten the United States now, if they ever did. Roughly a year later, the city she honored, Oklahoma City, Oklahoma, shook as a gigantic fertilizer bomb destroyed the federal office building there killing scores of people.

That spring of 1995 the international news media breathlessly covered the spread of the Ebola virus in Zaire. At the same time, in Washington, D.C., the U.S. Congress went ahead with plans to spend $60 billion on 30 new attack submarines while the Centers for Disease Control's budget for non-AIDS infectious diseases was set for $36 million. Scientists such as Donald Henderson of the Johns Hopkins School of Public Health warn that if the United States hopes to prevent a plague of Ebola or a similar virus from eventually sweeping the country, at least $150 million a year should be spent on early-warning clinics and virology labs in the field. This expenditure, never approved, is the equivalent of one-tenth the cost of a submarine, or one F-22 jet fighter. David Corn of *The Nation* (1995, p. 781) asks, "Which is the greater threat to the nation's security—the lethal virus or the unidentified enemy?"

In some human communities defining the real threats is not a problem. For example, for the people of Chiapas in southern Mexico, the threat is clear. It is the response that is complicated.

From Chiapas to the Future

> What governments should really fear is a communications expert.
>> —*Subcomandante Marcos to a* Newsweek *reporter (quoted in Watson et al., 1995, p 37)*

> Some say it is the first postmodern revolution, others say that it is the last Central American revolution, even geographically speaking.
>> —*Subcomandante Marcos on the Chiapas Conflict (Marcos, 1994, p. 10)*

Subcomandante Marcos traces the struggle he is part of to the Spanish invasion of Mexico, to "the Conquest that began, well, not exactly five hundred years ago, and that continues . . ." (Marcos, 1994, p. 10). But for all their long history there is also something new about the Zapatistas of the 1990s.

Theirs is a hybrid movement, with the traditional virtues of peasant rebellions augmented by media-savvy spokespeople who use the internet and the tabloid press with the shamelessness of athletic shoe companies. Marcos, the biggest Zapatista media star (but not the only one in Mexico by any means), is at times humble ("We are all Marcos." "I am merely a subcomandante.") and at others flirtatious. He ended one communiqué with "The Sup, rearranging his ski mask with macabre flirtatiousness" (Guillermoprieto, 1995, p. 44). He is also willing to kill and to die. *The New York Times* calls him "the region's first postmodern guerrilla hero" (Golden, 1994, p. A3). That may or may not be so, but he is clearly part of a sophisticated attempt by the Zapatistas to break their political isolation with a strange combination of local small unit attacks, national mobilizations, and international appeals.

Zapatista communiqués flash around the world on the Internet propelled by an electronic alliance of human rights and solidarity groups, appearing in Mexican newspapers and on hundreds of thousands of personal computers within hours of their release. The Zapatistas are very conscious users of "counterinformation" mainly distributed through the Internet.

Governmental responses have included the traditional killings and gang rapes, including at least one attack on a U.S. supporter traveling in Mexico, but they have also struck directly at the Zapatista infostructure, raiding the local purveyors of the Zapatista communiqués, arresting one as a Zapatista leader, and claiming that there is no way to check the accuracy of information on the Internet. But then how does one check the accuracy of the information in *The New York Times*? As it turns out, Internet information is much more quickly corrected and democratically collected than that in any print publication. Anyone participating in an ongoing newsgroup can easily keep track of what all interested readers, including those on the scene and with direct knowledge, think of what is posted. It is interactive information, constantly under scrutiny, unlike the information distributed by the multinational conglomerates who own the press and the rest of the mass media.

One of the major outlets for Zapatista thinking, *Zapatistas: Documents of the New Mexican Revolution,* was collected on the Internet and was first produced as an electronic text before being published by Autonomedia. In their communiqué of March 11, 1995, the Zapatistas specifically mention the role of "international civil society," mobilized in many cases electronically, in helping their cause. For a movement so focused on communication and information, an appreciation of the growing international electronic networks is not surprising.[3]

So how is it that this peasant/Indian rebellion mobilizes the most contemporary of weapons in the service of revolutionary objectives built around information and choice? This book is an attempt to explain this and the many apparent paradoxes of contemporary war. For the moment, it should be sufficient to give a long quotation from the articulate Subcomandante Marcos, who describes the goals of his 500-year-old struggle in terms that refuse the sureties of modernism, both reactionary and revolutionary:

> The people have to decide what proposal to accept, and it's the people who you have to convince that your opinion is correct. This will radically change the concept of revolution, of who the revolutionary class is, of what a revolutionary organization is. Now, the problem isn't in fighting against the other proposals, but instead in trying to convince the people. It's because of this that the Zapatista revolution isn't proposing the taking of power, it isn't proposing a homogeneous ideological concept of revolution. We are saying that yes, we do have our idea of how the country should be, but something is lacking before we talk about this. We cannot replicate the same logic as the government. They have a vision for the country that they have imposed on the people with the arms of the Federal Army. We cannot reverse this logic and say that now the Zapatista vision is going to be imposed on the people with the arms of the Zapatista army. We are saying, "Let's destroy this state, this state system. Let's open up this space and confront the people with ideas, not with weapons." This is why we propose democracy, freedom and justice. (Marcos, 1994, p. 11)

Victory, for Marcos, isn't achieving state power, it is reconfiguring power. When a *New Yorker* reporter told Marcos that it was a delusion to think the Zapatistas could really capture Mexico City, he replied, "Weren't we there already by January 2nd? We were everywhere, on the lips of everyone—in the subway, on the radio. And our flag was in the Zócalo" (Guillermoprieto, 1995, p. 41). At a Mexican press conference he noted, "We did not go to war on January 1 to kill or to have them kill us. We went to make ourselves heard" (Golden, 1994, p. A3).

*   *   *

Hitchhiking through Utah some 15 years ago I was picked up by a guy who told me to call him Red. He had just gotten off work, so we had a few beers and drove toward the setting sun. Red got to talking and told me of his job. He worked at the Moab uranium enrichment facility. A shift leader. Good pay, but . . . hard work. Lots of hassle regs. Sometimes he was so radioactive at the end of the day that even after ten showers he couldn't go home to his wife and daughter. Those nights he slept on site as he waited for his body to shed enough radioactivity from his skin, bowels, lungs, bladder, and mucous membranes so he could sleep with his family.

I have met many such people, each marked in their own way as deeply as Red by our war-ready world. I have seen many places that are part of the same system that runs Moab to make plutonium. At Torrejón, outside of Madrid, I lived with some of the young guys who load and unload supersonic fighter-bombers with nuclear explosives every day. In Barcelona I spent one New Year's Eve celebrating with U.S. Marines, who that same year found themselves targets in Lebanon. Inside Lawrence Livermore Laboratories I encountered a dozen of the scientists and managers who design these weapons and their delivery systems. In the backcountry of Vandenberg Air Force Base I have seen the test launchpads for intercontinental missile tests and spent many hours debating their morality with the people who improve them. Even now, in 1996, I have one student who is a Serbian writing about her war guilt and another who was a SEAL commando in Grenada, Panama, and the Persian Gulf. War is, and has always been, personal.

Coincidentally enough, my interests in war, and postmodern war itself,[4] both crystallized in the same place: Vietnam. I was a toddler when my parents took me to live in Saigon during a lull in the long Indochina War. There I collected old helmets and spent bullets with thoughtless pleasure. As I've grown older my feelings toward war have grown much more complicated and much less clear.

A few years ago I visited an embalmed Titan missile silo in Green Valley, Arizona. It floats on giant springs for riding out a near-miss. I sat in one of the easy chairs. I reached for the key. I pushed the red button. A year later I watched TV in horrified fascination as my country technologically dismembered Iraq, killing hundreds of thousands. Since then there have been many wars. Literally dozens.

It haunts me. The sheer weight of war's materiality and the violence of its inscription on the body politic, as well as my own body, force me to seek an explanation for the strange danse macabre of our age, war. The focus of my research, and what makes the current situation so particularly dangerous, is the increasing application of the incredible powers of technoscience[5] to war.

The central role of invention, science, and high technology in war today has many ramifications both subtle and obvious. The connection between war and technoscience has long been intimate; now it is integral. A certain basic level of technology is what makes war possible. But in the last few hundred years there has been an incredible increase in the interdependent connections between the forces of creation and of destruction. The latest, perhaps decisive, link is between computers and the military.

Current policy of the U.S. Department of Defense makes advanced computing a necessity: it is crucial for the military utilization of space; it is central to the AirLand Battle doctrine used in the Gulf War; it is a key part of drug wars and peacekeeping missions. In general it is integral to the C^4I^2 needs (command, control, communication, computers, intelligence, interoperability) of all types of war conceived of by the Pentagon, from "limited" (covert, guerrilla, net-, cyber-, counterterrorist) to "general" (nuclear, subnuclear, near-nuclear). This computerization of war is the culmination of a long process of mechanizing management and rationality in the U.S. military, and it has led to the system of postmodern war.

My particular theoretical approaches are varied and not even equivalent. They come from many disciplines (Foucault, 1980; Geertz, 1973; Latour, 1981) and form a "tool kit." They have been chosen because of their utility (at convincing, predicting, even intervening) and because of their beauty (ethically, aesthetically), not because any particular view is the *one* truth. I don't believe in one truth any more than the Zapatistas do. I don't expect a reader to say, "Ah, yes, that *is* the way it is." Instead, I hope for, "Oh, if you look at it like *that*, it does make some important things clearer, while making others obscure no doubt." To build something specific one uses particular tools. To use tools well involves knowing their strengths and weaknesses, their advantages and dangers. The choice of tools also determines the actual process of construction. In the case of this work my tools are chosen because I wish to learn something of the unfolding dynamics of postmodern war, and even play some small part in our practical understanding, and therefore shaping, of its discourse.

This desire is what leads to my interest in power and how it is generated. This study, then, is of the juncture of ideals, metals, chemicals, and people that makes weapons of computers and computers of weapons and soldiers. More, I want to understand something about the rules of this often deadly game, and about the rules for making those rules (the metarules). Gregory Bateson once remarked on the importance of this politically:

> The problem of the international game is how to change the rules, whereas game theory tends to give us solutions to the questions of how to not lose according to the rules as they now are. Nobody knows a thing about changing the rules of the game. (quoted in Brand, 1974, p. 27)

I have tried to focus on an analysis that makes a difference, rather than a view that claims to be truer than any other. This is also an implicit recognition that even if there were grounds for one totalizing vision it would be a mistake. Joseph Campbell, the mythologist, has set this stage clearly:

> Every claim to authority of the book on which pride of race, pride of communion, the illusion of special endowment, special privilege, and divine favor were based has been exploded. . . . The faith in Scripture of the Middle Ages, faith in reason of the Enlightenment, faith in science of modern Philistia, belongs equally today to those alone who have as yet no idea of how mysterious really, is the mystery even of themselves. (1962, p. 609)

This collapse of totality can be embraced even if the confusion that has resulted sometimes seems like a suffocating void. A poet (Ractor, 1982) has written, "She whispers fantasizing: The Chamber is barren. All of us recognize our void view."

Going beyond "our void view" is not easy. It depends on where we choose to look. Donna Haraway, the philosopher of science, points out one possible direction (1985). She revels in how modern technoscience is producing incredible changes in ourselves and on our actual bodies as well as in and on the world around us. A central symbol of this is the cyborg—cybernetic organism—a human integrated with a machine. That the cyborg is coming true on the battlefield and that he[6] is a central metaphor of discourse about future war is an important theme of this book.

For Haraway, the way to deal with this cyborgization of ourselves is not back, or around, but through. In her essay, ironically titled "A Manifesto for Cyborgs," Haraway insists that

> taking responsibility for the social relations of science and technology means refusing an anti-science metaphysics, a demonology of technology, and so means embracing the skillful task of reconstructing the boundaries of daily life, in partial connection with others, in communication with all of our parts. It is not just that science and technology are possible means of great human satisfaction, as well as a matrix of complex dominations. Cyborg imagery can suggest a way out of the maze of dualisms in which we have explained our bodies and our tools to ourselves. This is a dream not of a common language, but of a powerful infidel heteroglossia. (1985, p. 100)

This work is of that dream.

*   *   *

Last night I went to a college basketball game with my family. Before the game the band played "The Star Spangled Banner" and we sang along. It took me a few moments to notice that the flag I was hailing was an electronic image on the scoreboard—a virtual flag waving in a simulated breeze under electronic sunlight. But looking around at the gathered Oregonians I could tell that their feelings for that simulated flag, and whatever it represents, were real. False flags can inspire real fervor. Doctrines of war can certainly breed real conflicts, as can myths, metaphors, and illusions. With war, people are always in denial. But under the many layers of talk about war, from the art of war to infowar, there is war itself, feeding upon dead and maimed bodies. To deny this is just to ask for the return of what it is that is repressed—the bloody reality of war.

This book is written against denial. Denial takes many forms. It can claim that war is mere spectacle and simulation, as some postmodernist theorists do. It can claim an end of history, as many conservatives have. It is the infatuation with new superficial theories of pseudowar, such as cyberwar, in the face of the apocalyptic dangers of real war. It is to assume that real peace is not possible. All of these denials could prove fatal to humanity for war is not just interested in us now, as Tolstoy and Trotsky apparently warned. It is more than interested. War has us in its grip, and we have it. We shall determine our future, if any, together.

Part One

The Present

Metropolis collage, by the author.

Only a test: Israeli students wear gas masks to class Wednesday during a nationwide security drill during which mock chemical warfare conditions were conducted in civilian situations.

Exhumation detail: A Serb steals the casket of an exhumed body from a Serb-held cemetery Wednesday in Ilidza near Sarajevo. Many Serbs are digging up relatives in preparation for fleeing the area.

A day in the life of postmodern war. Four pictures from the January 25, 1996, Oregonian, page A2. Mideast, Balkans, Asia, Central America. As the twentieth century comes to an end various conflicts fester and rage on every inhabited continent. They may range from domestic terrorism

Peaceful stroll: An Indian girl walks past a Mayan cross and a miltary police officer outside Mexican–Zapatista peace talks in San Andres Larranzar. Rebel leaders said Wednesday that they would sign an agreement expanding the rights of Indians clearing the way for a peace accord in Chiapas state.

Upset over islet: Protesters burn a Japanese flag Wednesday near the Japanese Embassy to dispute Japan's claim that it owns tiny Tok-do, an island halfway between South Korea and Japan.

to conventional war but all carry the potential of apocalypse, thanks to the creation and proliferation of biological, chemical, and nuclear technologies and weapons. Paradoxically enough, all of these conflicts seem poised on the edge of peace. Postmodern war's incredible instability is both a tremendous danger and a utopian possibility. Photos: AP Wide World Press.

[15]

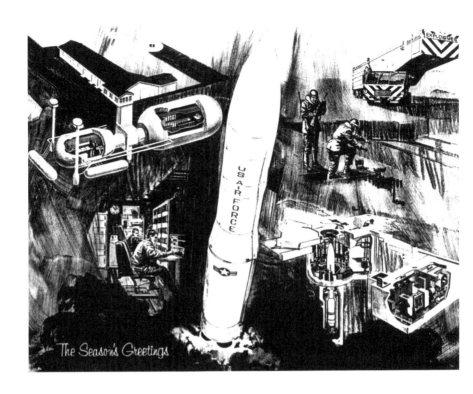

The Season's Greetings

The normalization of unspeakable horror. A Christmas Card from the Minuteman Service News, *the journal for the ICBM crews of the U.S. Air Force. While many nuclear warheads and ICBMs have been dismantled, enough remain to destroy the world as we know it. So many Christmases have passed under this sword of apocalypse that few now fear it. That is not a rational reaction, but then making an ICBM the centerpiece of a Christmas card isn't either. Yet it is a peculiar version of official rationality, in technology, science, and politics, that has justified, and therefore actualized, these weapons.* Minuteman Service News, *November–December 1974, no. 77, page 3.*

The imperialistic hubris of official icons. The world is their oyster. While all official insignia from the Department of Defense do not inscribe the world as militarized U.S. territory it is remarkable just how many do. Here we see the badges for commandos (the Navy SEALs and the Joint Special Operations Command that unites all special forces), for space forces (Space Command), for researchers (the Strategic Computing Program and for DARPA as a whole) all feature the world as their backdrop. The other logos are corporate (GTE) and from corporate–military conferences in the Czech Republic and the United States. The message is clear in all of these symbols. War is global, militarization is global, our reach is global.

Chapter One

Real Cyberwar

> Resistance to Pure War can only be based on the latest
> information.
> —*Paul Virilio and Sylvere Lotringer (1983, p. 136)*[1]

Simulation and Genocide

During the Bosnian peace talks of 1995 the negotiators for the United States
used state-of-the-art real-time simulation computer technoscience to bring
a temporary halt to the genocide. At a key point in the Dayton Accord
meetings U.S. officials took the Bosnian, Croatian, and Serbian presidents
into the "Nintendo" room where they could see a real-time three-dimensional map of the disputed territories. Secretary of State Warren Christopher
bragged that "We were able to, in effect, 'fly' the people over the area they
were talking about, showing them the map on a large video screen so they
could actually see what they were talking about." This is why the negotiations
were held at Wright–Patterson Air Force Base in the first place; it is where
the term "bionics" was coined (Steele, 1995), where science fiction writers
are invited to plan for future war (C. H. Gray, 1994), and where virtual reality
and human–machine interface applications are developed. The French,
Russians, and other Europeans excluded from this show were, diplomatically
speaking, pissed off. Still, it worked, even though, as the French noted, the
map did not include data on the ethnic origin of the population.[2] The promise
of total information, which for many today means total power, was enough
to convince the warring presidents.

This incident was just one of literally millions of war-related events that
happen every day. But its mixture of ancient hatreds with the newest of
technology in the service of intimidation, maybe even peacemaking, at the
behest of the world's "only" superpower[3] make it a good one to unpack. It
reveals, among other things, how powerful, but limited, the forces pushing
for peace and stability are. The international model of open war now equates

[19]

war to a virulent disease, when it does not see it as a righteous cause, which it usually does. So while everyone calls for peace, every conceivable justification is mobilized for war, including God, blood, gold, honor, oil, water, history, and the need for peace itself. And many who make these justifications are even sincere, none more so then those who used the latest computer technology to attempt the inoculation of potential combatants against the seductions of more genocide in blood-soaked Bosnia.

The Bosnian conflict, in which satellites are used to find the mass graves of the recent victims of thousand-year-old hatreds, isn't the only confusing conflict. Consider these:

- PSYOPs (psychological operations) U.S. troops battle voodoo imagery in Haiti.
- Russian troops attempt with little success to crush cell-phone-linked Moslem rebels in the Caucasus.
- Religious revolutionaries attack liberal regimes with assassination (Algeria, Egypt), bombs (United States, France), and gas (Japan), even in alliance with their hated enemies, as with the Orthodox Jewish and Hamas Moslem alliance in Israel against peace. On June 15, 1995, President Clinton orders the United States to prepare for bioterrorism (network TV reports, January 30, 1996).
- Zapatista insurrectionists shake Mexico using a combination of traditional and cyber-guerrilla tactics and strategies.

But this bricolage, this . . . mess, is not unexplainable. It is just that the explanation isn't simple. Imagine war as Proteus, the Greek demigod who could change his form as long as he touched the ground. Hercules had to hold him up in the air during their wrestling match in order to defeat him. Well, protean war is wrestling with the present, trying to find a way to survive as a coherent creature (discourse, culture, or way of life, the label can vary) in the face of extraordinary changes in the technologies and politics of conflict.

War explodes around the planet, relentlessly seeking expression in the face of widespread moral, political, and even military censorship, since the old stories of ancient tribal grievances and of the supremacy of male courage, and therefore war, don't sell everywhere. Peace even seems particularly popular and much effort, much military effort, goes into what is labeled peacemaking, but the situation remains chaotic. Continual technoscientific changes have led to the incredible spread of mass weapons (chemical, biological, nuclear [CBN]) to dozens of states, nationalities, and even grouplets, while the continual outbreaks of war are almost contained by the spread of worldwide high-speed communications, the integration of the world economy, and the proliferation of peace initiatives.

Just what is happening? It is the argument of this book that war is undergoing a crisis that will lead to a radical redefinition of war itself, and that this is part of the general worldwide crisis of postmodernity. While this crisis is far from resolved, the range of possible outcomes is becoming clear. In terms of war there would seem to be two main future options: a utopian redefinition of war's function (simulation), or some sort of horrific, even apocalyptic, *reductio ad absurdum* of real war's current logic (genocide).

To sort out the whys and the maybes of these possible futures a great and complicated story needs to be told—the story of postmodern war. It starts with a discussion of the label itself.

Why "Postmodern" War?

The mania of the last few hundred years for labeling new types of war has a sound basis. War seems to be changing more quickly then it ever has before, while it assumes unprecedented destructive power as well. Labels can be misleading, as I will show in the case of "cyberwar" below, but they are also important. I don't choose mine lightly. Whether, for example, the Persian Gulf War is seen as a new form of war or as a continuation of older types is more than an academic issue.

"The Persian Gulf conflict," intones *Business Week*, "will almost certainly come to mark the transition between two forms of war." The magazine goes on to specify what it considers the most important elements of this "new war": (1) "the integration of high-technology-based systems"; (2) "a huge array of computer and communications systems"; and (3) the realization that "politics and public relations play" a crucial role "in achieving military objectives" (*Business Week* Staff, 1991a, b, c, pp. 39, 42, 37).

But this is not new. It is not all that different from Vietnam. While the War for Kuwait certainly broke some new ground (and most violently), for the last half of this century it has been clear that war is changing fundamentally. The more insightful observers have noticed the implications of high-tech weapons, especially computers, and the permanent military mobilization that has existed since 1945. They have called this new type of war many things, among them: permanent war (Melman, 1974), technology war (Possony and Pournelle, 1970), high-technology war or technological war (Edwards, 1986c), technowar or perfect war (Gibson, 1986), imaginary war (Kaldor, 1987), computer war (Van Creveld, 1989), war without end (Klare, 1972), Militarism USA (Donovan, 1970), light war (Virilio, 1990), cyberwar (Davies, 1987; Der Derian, 1991; Arquilla and Ronfeldt, 1993), high modern war (Der Derian, 1991), hypermodern war (Haraway, personal communication, 1991), hyperreal war (Bey, 1995), information war, netwar (Arquilla and Ronfeldt, 1993), neocortical warfare (Szafranski, 1995), Third Wave

War (Toffler and Toffler, 1993); Sixth Generation War (Bowdish, 1995), Fourth Epoch War (Bunker, 1995), and pure war (Virilio and Lotringer, 1983).

Though all of these labels have something to recommend them, none do justice to the complexity and sweeping nature of war's recent changes. For example, Virilio and Lotringer's "pure war" does capture poetically the deep penetration of war into contemporary culture, certainly in the West, especially into politics.[4] But in a strong sense the mass annihilation of civilians in World War II was pure war, the obliteration of Hamburg, Dresden, Tokyo, Hiroshima, and Nagasaki countering that of Warsaw, Rotterdam, and Coventry, not to speak of the earlier "Rape of Nanking" and the "Final Solution" of Treblinka, Auschwitz, Dachau, and the other death camps. War was total. What we have now is very "impure" war, called "imperfect" war by some, coming to the fore because pure total war has become, thanks to technoscience, suicidal. War is diffused throughout the culture, helping shift gender definitions, structuring the economy, selling products, electing presidents, and boosting ratings. But the actual battles are not decisive or heroic; they are confusing, distant, and squalid. More is happening to war than just high technology, or computers, or speed as light, or cybernetics, or the militarization of information.

I call it postmodern war.[5] Why choose "postmodern" over the other possible labels? There seem to be two good reasons. First, modern war as a category is used by most military historians, who usually see it as starting in the 1500s and continuing into the middle or late twentieth century. It is clear that the logic and culture of modern war changed significantly during World War II. The new kind of war, while related to modern war, is different enough to deserve the appellation "postmodern." Second, while postmodern is a very complex and contradictory term, and even though it is applied to various fields in wildly uneven ways temporally and intellectually, there is enough similarity between the different descriptions of postmodern phenomena specifically and postmodernity in general to persuade me that there is something systematic happening in areas as diverse as art, literature, economics, philosophy, and war.

This is particularly true of the importance of information to postmodernity. As a weapon, as a myth, as a metaphor, as a force multiplier, as an edge, as a trope, as a factor, and as an asset, information (and its handmaidens—computers to process it, multimedia to spread it, systems to represent it) has become the central sign of postmodernity. In war information (often called intelligence) has always been important. Now it is the single most significant military factor, but still hardly the only one.

We have been living in the era of postmodern war since 1945 and cold wars are an integral part of it. Edward Luttwak defines *cold war* as: "International conflict in which all means other than overt military force are used"

(1971, p. 244). The United States versus USSR "Cold War" (capitalized) was the most famous of these, but it wasn't unique. Calling cold war (lowercase) info- or cyberwar doesn't change it significantly.

Today, all confrontations between the nuclear powers are constrained from total war by the threat of nuclear weapons and other devastating technologies. The tensions between the United States and any other Great Power are relegated to low-intensity conflicts with proxies, political struggles, and economic competition. Postmodern war depends on international tension and the resulting arms race that keeps weapons' development at a maximum and actual military combat between major powers at a minimum. Real wars still happen, sometimes despite what the Great Powers wish. High-intensity, large-scale war could always break out, which in part explains the growing efforts to constrain violent conflicts. But continuing illusions about the nature of war itself bedevil any attempts to bring it under control. War cannot be managed: it is not a science; it cannot be controlled. But the seductive logic of control theories is incredible. The latest batch of cyberwar and information theories are particularly sexy—and dangerously limited.

Cyberwar, the Next Step?

Information war must be the hot new thing. It's been on the cover of *Time*, Newt Gingrich has given a speech about it, Jackson Browne has a song called "Information Wars," Tom Clancy has a novel, and RAND has a big research review.[6] "Cyberwar is Coming!" exclaims the title of John Arquilla and David Rondfeldt's influential article from 1993. Not the kind of staid academic warning one expects from analysts in RAND's International Policy Department or from political science journals. The article had little in it that hadn't been mulled over by non-RAND academics, science fiction authors, and middle-level military officers who have been worrying about the impact of the information revolution on war since the early 1980s, and it wasn't even much different from certain currents of military thinking that can be traced through the whole history of war back to Sun Tzu's *The Art of War*. But it has certainly caught on, spawning dozens of similar articles and coverage in the mainstream press.

There are three distinct elements of cyberwar that its proponents see as crucial.[7] First, there is the illusion that war can be managed scientifically. This is a very old idea that goes back at least to the 1500s. Second, there is the realization that war is in large part a matter of information and its interpretation, especially as politics, and that it is in this realm that the metarules of war are set that determine who wins. This is not a new concept. It can be found in *The Art of War*, and it is the central insight of the whole range of guerrilla ("little") wars that includes ragged, colonial, irregular, and counterinsurgency wars and low-intensity, cold wars as well as the vast area covered by operations other than war (OOTW).[8] And, third, there is an emphasis on the importance of computers and how computers change traditional limits of war in terms of time and space.[9]

This last area truly is new and significant, although the contemporary military theorists have not gone far beyond the wet dreams of electronic battlefield theorists of the Vietnam era except by theorizing that cyberspace itself is a battleground. That claim has yet to be proven in combat, although the theorists do note that it is the most computerized army that is the most suseptible to netwar—which means the U.S. military.

Pentagon doctrines and cyberwar games paint this grim picture in lurid colors. According to *The Washington Post* over 95 percent of U.S. military communications go over civilian networks. There are at least 150,000 military computers tied directly to the Internet. In 1994 a team of in-house hackers was unleashed by the Defense Information Systems Agency on military computers and penetrated 88 percent of the nearly 9,000 attacked. Only 4 percent of the penetrations were even noticed. There were 350 real hack attacks noted in 1994; therefore, an estimated 300,000 penetrations were made into U.S. military computers in 1994 if 4 percent were noted.[10]

In cyberwar games staged in 1995 the enemy (Islamic fundamentalists hiring Euro-hackers) used software viruses, worms, and Trojan horses to crash trains, planes, and banks before bringing the United States to its knees by cutting off phone service. As one participant remarked, "This was not something that carpet bombing was going to solve."[11] Of course, carpet bombing hasn't ever solved any geopolitical problem (Sherry, 1987; Clodfelter, 1994), but the myth that military technology solves political problems is a strong one. So the enthusiasm for cyberwar is unabated and the United States is rushing to develop the traditional doctrinal and bureaucratic infrastructure with which it responds to every expensive new revolution in military affairs (RMA).

Every branch of the armed services has its take on cyberwar,[12] and so do the Canadians and at least three dozen other countries. While some old soldiers openly attack "half-baked ideas from people who have never been shot at," the general trend is clear. As the U.S. military is forced to avoid the risk of any casualties, the promises of bloodless cyberwar are all the more seductive.[13]

This leads inevitably to the more flexible definitions of war that make of every case of domestic dissent and criticism an act of low-intensity conflict or "netwar."[14]

Many nongovernmental political groups see the same terrain as netwar. The Critical Art Ensemble[15] declares:

> The rules of cultural and political resistance have dramatically changed. The revolution in technology brought about by the rapid development of

the computer and video has created a new geography of power relations in the first world that could only be imagined as little as twenty years ago: people are reduced to data, surveillance occurs on a global scale, minds are melded to screenal reality, and an authoritarian power emerges that thrives on absence. The new geography is a virtual geography, and the core of political and cultural resistance must assert itself in this electronic space.

The Critical Art Ensemble doesn't respond in a military mode; however, it seeks to change the way people relate to the infosphere through, for lack of a better term, art. Other net activists are not so subtle. Jason Wehling buys into the infowar model with enthusiasm, proclaiming "the attack has already begun." He advocates various disruptive and confrontational tactics to go along with the continued spread of information and building of international solidarity that has proven so effective for the Zapatistas (Wehling, 1995).

It is Hakim Bey (1995) who has done the best job of integrating the insights of cyberwar with the realities of genocide. For all its mumbo jumbo about pure information, the Terminal State still needs a military and police force that is unprecedented in size and power. War is, and always must be, not just for the hearts and minds of people but for their bodies as well. Demonizing, or deifying, information won't change this, nor does it help. As Bey notes, "information is a mess"; it is what we do with our living bodies in terms of that mess that determines politics. This is clear everywhere contemporary war continues.

The Experience of Contemporary War

The first dead U.S. soldier in Bosnia was killed by a land mine. President Bill Clinton was not entirely inaccurate when he said, "he died in the noblest of causes—the pursuit of peace."[16] But it is hardly one that is guaranteed to succeed. Still, if high technology is the key to victory, the international forces certainly have a chance. Not only do they deploy the latest weapons and weapon platforms (tanks, helicopters, planes), but the informational infrastructure is state of the art. Tactical information systems linking radar and gun control can pick up every shell fired and trace its course while communicating to users with sophisticated multimedia interfaces. These systems, such as the French designed Safari from Alcatel, are designed to share many different types of information in human and electronic languages.[17] Interoperability and evolution are two of the system's strength, as is its experience in Somalia, Rwanda, and Bosnia.

The theory behind the technology deployed in Bosnia, for example, is pure cyberwar. At a 1995 conference in the Czech Republic to introduce the

Czechs and visiting Poles to NATO's new infowar doctrines, there was total acceptance. A Dutch naval officer spoke in cyborg terms about the role of computer technology in creating a human–machine "fighting organism." A representative from Thompson's stressed that it all depends on having higher levels of human–machine interfacing and declared "Winning means having the best information." Czech and Polish General Staff officers elaborated on their own infowar infrastructure projects.[18] In terms of peacemaking or peacekeeping, this approach may well make sense, as the very ground for peacemaking is communication.

Peace activists use a similar analytic and have put a great deal of energy into creating internet connections between the former Yugoslav component republics and the rest of the world. The ZaMir network they established has proven to be not only resilient but very effective in keeping communication open.[19] But projects such as this sometimes seem a brief ray of sunshine in a storm of war.

As the millennium plunges into the second half of its last decade virulent wars sputter on despite massive international attempts to bring peace. In Ireland, in the Middle East, in the Balkans, and in Central America cease fires and peace treaties follow massacres and battles with relentless regularity. There is a self-perpetuating cycle of revenge in all these places, as in Israel,[20] and in Northern Ireland, where renewed bombings led to increased British militarization in early 1996.[21] Still the balance of power seems to be slowly tipping toward peace. It seems probable that some, if not all, of these wars will be resolved before the year 2000, although the Middle East has the potential of going completely the other way into apocalypse.

In all of them conflict will certainly continue, but perhaps below the physical violence threshold. In all of them the marginalization of violence is ongoing. In Bosnia it is constrained by the direct threat of massive NATO violence, backed up by the all-seeing eye of the milsat (military satellite) panopticon. In Ireland and the Middle East it is only the rejectionists (from the security apparatus as well as the religious fundamentalists) that continue to fight, although there are many of them. In Chiapas, Mexico all sides have been very careful to constrain the violence.

One could say that the United States is at netwar with Chiapas, Mexico. In a "war" that only included a few weeks of violence, the Zapatista rebels may have managed to secure at least one victory. The February 1996 peace pact expands the civil rights of Mexico's Indian people, and old massacres are to be investigated anew.[22] And with direct violence minimized, the survival potential of the Zapatistas is high. Unfortunately, the underlying problems, (ecological crisis, inefficient autocracy, contradictions of capitalism, ferocity of the world economy) remain the same. So it seems for many contemporary crises. The problems fester.

Benjamin Barber argues that contemporary democratic politics are being torn apart by "Jihad" and "McWorld" (1995): on the one hand, the promise of consumption (McWorld) draws many into the world economy as consumers; on the other, some are frightened into a fundamentalist reaction (Jihad). Liberal nation-state-based democracy is eroded from both sides. Expectations, material and spiritual, seem to be spiraling out of control. Can everyone have enough to consume? Can all nations rule all the land they ever occupied all at once? Can we be pure? These questions are being answered in blood around the world, from Central Africa to Timor and all along the edges of the old Soviet Empire. It will get worse before it gets better, especially in the Third World, the traditional killing ground of postmodern war.

High-Tech, Low-Intensity Deadly Conflict

> The Cold War was red hot in the Third World.
> —*John Brown Childs* (1991, p. 82)

Since World War II most conflicts have been in the Third World. Childs (1991) points out that racism is a significant aspect of this and that

> the Cold War was fought with the blood of Third World peoples both allies and opponents. From Angola and Mozambique, through Guatemala, El Salvador, and Nicaragua to Chile, the Philippines, Ethiopia, Somalia, Zaire, Lebanon, the Dominican Republic, to name but a few (p. 82)

Usually, the Great Powers had "setbacks" in these wars, although the system of United States–USSR hegemony survived, but it was other people who were defeated, sometimes fighting for themselves, sometimes just fighting as proxies. The proxy relationship is sophisticated. It should be. It is as old as empire. These days the Great Power may well supply weapons, advisers, training, strategy consultation, and so on. The value of the best high-tech weapons systems is high. The actual war is used to drain the other superpower and to test new weapon systems, and usually to justify domestic repression as well. This political–ideological management of world and domestic order was much more effective than direct military intervention. For once large numbers of Soviet or U.S. troops became involved real defeat was possible, and it came, throwing the system into its postmodern crisis.

The faith in high-tech war is still high. Even the superpower defeats in Vietnam and Afghanistan left the believers in technoscience unfazed: "It was not so much the irrelevance of high technology, but its misapplication— either in tactical, operational, strategic, or political terms—that contributed to the failures in Indochina" (Zakheim, 1988, p. 236). Dov Zakheim has several other arguments. First, "it is an irrefutable fact" that non-Western countries are buying their own "arsenals of sophisticated weapons." This "proliferation of high-tech weaponry . . . has ratcheted up the requirement" for the technological powers which might seek "to project their forces into these [Third World] areas." He concludes that "high technology is central to most potential battlefields of the Third World" (p. 236). But is that true?

Any serious consideration of U.S. low-intensity conflict (LIC) policy has to start with operational reality. The United States is almost always waging or supporting a number of bloody low-intensity conflicts. Although much of this activity is still secret there are cases about which a fair amount is known, such as Nicaraguan contra support, the Lebanon interventions, Afghan proinsurgency assistance, and the ongoing U.S.-financed fighting in El Salvador and Guatemala. Other recent LICs include the naval operations in the Gulf of Sidra and the Persian Gulf, and the invasions of Grenada and Panama. Recently, the trend has been for humanitarian missions, such as Somalia, Haiti, and Bosnia.

The covert budget for LIC was $2 billion in 1987. Add to that the over $2 billion in acknowledged funding and hundreds of millions to buy high-tech weapons and equipment, including Polaris nuclear ballistic submarines modified to carry commandos, laser-guided weapons, and an incredible array of complex electronic sensor and communications equipment, and its clear that over $5 billion was spent on low-intensity war by the U.S. Department of Defense and the intelligence agencies in 1987 (Weiner, 1991). The figure has certainly gone up in the nineties.

There are more then 20,000 covert operations soldiers under Pentagon control now, including Army Green Berets and Rangers, Navy SEALs, and various Marine and Air Force commandos. Together they make up the Special Operations Forces. Their motto is "Anything, Anytime, Anywhere, Anyhow."

The Air Force's related policy slogan to justify an air–space role in LIC: "Global Reach is Global Power." But this is more than a rhetorical move for budget priorities. Lt. Gen. Carl Stiner, on-site commander of the invasion of Panama, has even claimed, "I can't go to war without space systems" (*Military Space* Staff, 1990d, p. 3).

Indeed, there is evidence to support the cheerleading claim of the *Military Space* newsletter that "Space systems played an important role in [the] Panama Invasion." Among other systems the Operation Just Cause U.S. invasion forces used the Defense Meteorological Satellites, the Defense Satellite Communication System, ultrahigh-frequency (UHF) and super-high-frequency (SHF) satellites, the Global Positioning System (for helicopters and ground troops), and even civilian Landsat remote sensing satellites (*Military Space* Staff, 1990e, p. 8).

Space systems are just part of the technophilia of LIC. Complex radios, code computers, speedboats, stealth air vehicles, explosives and sensors, remote controlled drones and missiles, are all part of LIC practice. During the contra war against Nicaragua, the United States supplied them with a dozen KL-43 encryption machines, seismic, acoustic, magnetic, and infrared remote sensors, lists of satellite-generated targets, remote piloted vehicle (RPV) flyovers, and other hardware (M. Miller, 1988, p. 21). Since 1980, most U.S. commando teams have maintained satellite uplinks by using UHF Manpack terminals that weigh around 18 pounds. New versions with much more power are due out in 1998.

The elite units are a particularly enthusiastic military audience for new technologies. One industry satcom (satellite communications) expert says, "They really love gadgets." Or, as the trade journal *Military Space* puts it, "In addition to doctrinal flexibility, elite units also are willing to try new system concepts" (1990f, p. 1).

In many ways, though, it is the new system concepts that are trying them, and this is just one of the elemental ironies of postmodern war, which are legion—the most poignant being the persistence of war after the U.S. Cold War victory.

The Policies of Pax Americana

For those that predicted that peace would break out around the world after the collapse of the USSR consider the comments Alvin Bernstein of the National Defense University made in 1990 to a meeting of historians in Maryland. Speaking as a replacement for a high defense department official, Bernstein made a strong argument for high-tech weapons to "deter and contain regional conflicts." He especially liked "stand-off ordnance" that wouldn't involve risking U.S. soldiers. Korea, India, the Persian Gulf, and unspecified "terrorist states" were mentioned as possible loci for these regional conflicts. But Bernstein still couldn't let go of his desire for the United States to keep large ground forces. In a revealing comment, he argued that land armies should be retained in this post-Cold War period of threatening budget cuts to protect the United States against a *"currently anonymous and certainly ineffable enemy."*[23] Now, "ineffable" has some interesting meanings, most notably: "too overwhelming to be expressed or described" and "too awesome or sacred to be spoken." Within weeks of Bernstein's comments, this sacred role was being fulfilled by Iraq and the Pentagon budget was safe.

But it is not that simple. It seems Lebow and Stein (1994) are right when they argue that *We All Lost the Cold War*. War won. A state of war is still deemed necessary. For the United States, the Soviet menace used to supply the enemy and the Cold War was ongoing. Now the enemy is more diffuse, although just as ubiquitous. Domestic communists have been replaced by drug pushers and users, international communist agents by international terrorists, and the Cold War by perpetual low-intensity wars, with an occasional midintensity conflict as well. The persistence of war is clear.

From the realization that the domestic culture is the main problem in winning LICs to militarily attacking part of that culture may seem like a small step, but for the U.S. military it was actually a giant leap—a leap it was not prepared to make until the collapse of the Soviet threat. For years President Ronald Reagan tried, with very little success, to get the military to commit itself to the drug wars. It took Mikhail Gorbachev to make it happen.

There was a very gradual increase in the military's involvement. It started with electronic and other intelligence gathering. Then the Navy and the Air Force started helping with interceptions of smugglers. At first only the border areas were militarized, ostensibly to prevent the infiltration of terrorists and the importation of drugs. As early as 1986 the Marines were running clandestine electronic surveillance on the Mexican border. This militarization of the border actually serves many functions. Generally it increases the crisis mentality of the wars on drugs and terrorism. It also puts pressure on Mexico and on illegal immigrants, useful for various economic reasons such as keeping illegal aliens from organizing and demanding fair treatment.

The policy of militarizing the border psychologically reached its most absurd rhetorical epiphany when Reagan warned that Nicaraguan terrorists were only a two-hour drive from Harlingen, Texas. While less amusing in its rhetoric, the policy of the Bush administration was similar in most respects (Moore and Sciacchitano, 1986, pp. 8–9), and President Clinton has continued in this vein, even appointing a general as drug czar.

Infantry armed with automatic weapons have lately been sent sweeping through small towns and rural neighborhoods in Northern California, and other troops have been deployed at rock concerts to help the police search the crowd for drug users. The U.S. Air Force Academy has even sponsored a whole conference on how satellites and other space assets can be used in the war on drugs. By the spring of 1990, the military was asking for over $1.2 billion for the war and the Navy was planning to deploy carrier battle groups and even Trident nuclear missile submarines on antidrug missions. Vice Adm. Roger Bacon went so far as to claim that Tridents (which only carry strategic nuclear missiles) are "essential as . . . a defense against terrorism, drug trading and other global conflicts" (Morrison, 1990, p. 3).

A turning point was reached in the summer of 1990. It was then that the military involvement in the drug wars escalated, almost certainly as a reflection of Pentagon fears of a peace dividend. Regular troops were deployed for the first time under the direct approval of Secretary of Defense Richard Cheney. Two hundred soldiers of the 7th Infantry Light Division were sent to the Emerald Triangle in Northern California to search public lands for marijuana plants. In the process they confronted more than a few Californians along forest trails or in their own backyards. Armed with M-16 automatic weapons and supported by transports and helicopters, they whipped up so much anger that within a few weeks the *San Francisco Harold Examiner*, the flagship paper of the Hearst chain, called for the legalization of drugs and an end to the drug "civil" war.

That same summer more than $45 million was spent by the Department of Defense (DoD) to facilitate over 3,500 National Guardsmen and 65 Huey helicopters joining in the drug war. The National Guard was active in the drug war in all 50 states for the first time. In Kentucky, citizens were encouraged to report on "any suspicious activity," including any "house in the country where men are constantly going in and out" (Isokoff, 1990a). Forty Massachusetts National Guardsmen, supported by a helicopter, and approved directly by the Pentagon, searched the patrons at a Grateful Dead concert in Foxboro and fingered over a hundred people for the police. Forty arrests were made (Isokoff, 1990b).

The military also began to use RPVs in the drug war, specifically on the border. The 1st RPV Company of the U.S. Marines spent three weeks on the Texas–Mexico border and helped capture 372 illegal aliens and seize 1,000 pounds of marijuana. This was the first time the FAA had allowed "free" (without escort) RPV operations in civilian air space. The U.S. Army was testing RPVs off Florida at the same time, and the Drug Enforcement Agency (DEA) was investigating the value of various UAVs (unmanned air vehicles) in "following suspects, border patrols, and crop hunting." The Immigration and Naturalization Service (INS) and FBI are developing autonomous mechanical border guards and other types of narcotics officers as well (Lovece, 1991, p. 47).

A drug war institutional infrastructure is growing throughout the military, as it has already spread in law enforcement. It becomes one more institutional argument for continued expansion of the drug wars. Training centers, special units, and ground facilities are the sites of great bureaucratic, military, and rhetorical power. For example, the National Guard runs a training school, the National Counter-narcotics Institute. Due regard is given to the current intense bureaucratization of war. "The program features a three-day classroom 'war games' exercise in which potential drug war scenarios were resolved through coordinating different government agencies." In 1990, the DoD committed to a two-year $2 billion antidrug battle to expand Caribbean surveillance and to build a radar network from California to Florida (Myles, 1991, p. 15).

Drug wars interpenetrate with other types of war, the best example being the invasion of Panama. Here was a military "arrest" at the cost of at least 300 civilians dead and several dozen U.S. soldiers as well, most killed by friendly fire. It had to be the largest set-piece battle in a drug war since the British fought China to keep the market for opium open. In reality it seems the invasion of Panama was aimed more at charming the domestic media, attacking the most leftist neighborhoods in Panama City, and disciplining a hireling (Manuel Noriega) who had forgotten his place. Drug wars also play a very important part in justifying other conflicts.

For many countries, the bulk of U.S. aid is military, and much of that is set aside for drug wars. In 1991, such dedicated funds that were publicly known about included those to Peru, $36 million, Colombia, $76.2 million, and Bolivia, $53.2 million. Such aid may well encourage the further militarization of these countries and undermine what little democracy they have. For example, U.S. Special Forces units planned and led the assassination of at least one Colombian drug lord (Royce and Eisner, 1990). In Peru the militarization of drug wars has driven drug lords into alliances with rebels. The military uses helicopters and Green Beret-trained Peruvian antidrug police to stage massive raids. Herbicides are sprayed on suspect drug fields. There are U.S. troops in both Peru and Bolivia (Renique, 1990). In Guatemala, U.S. policy is to blame Nicaragua's Sandinistas and the FMLN (Frente Militar de Liberación Nacional) in El Salvador, as well as local rebels, for drugs and to cover up the deep involvement of Guatemala's secret police and military in such traffic.[24]

The drug wars are a series of related conflicts, including the efforts of various states to repress the production, sale, and use of certain drugs. These same governments support other drugs, sometimes the most harmful (alcohol, nicotine), and often the secret agencies of these governments even facilitate the production and sale of drugs their own governments have labeled illegal. There is also the struggle of drug suppliers to seize some level of political power, as in Colombia, and there are the struggles between suppliers for vertical and horizontal monopolies. Not less than 400,000 U.S. residents are arrested for using or selling small amounts of drugs. Thousands more are killed, most in the bloody clashes between "entrepreneurs" in the unorganized capitalism of the illegal drug economy. The drug wars are particularly personal. Almost any American may suddenly become a civilian casualty, shot down in a crossfire, robbed by an addict feeding an expensive habit, or captured by the authorities for using a proscribed intoxicant, as opposed to one of the many legal ones. Is it surprising that on the right as well as the left many people feel that their government is at war with them?

Still, no matter how ubiquitous, the drug wars are not quite militarized enough for the real military, except perhaps for the arrest of Noriega. They are being incorporated into the system of war in a very new way, but only as part of the struggle against a larger enemy—terrorism.

The terrorist threat is also particularly personal: businesses, tourists, airline passengers, and government workers are all potential targets. Thus there has been substantial growth in the private security field, linked again with the drug wars. In 1989, the market for "security services and related paraphernalia" was somewhere between $20 billion and $50 billion annually. One estimate was that there were more than 5,000 companies engaged in the security business in the United States by 1990 (Livingstone, 1990, pp. 36–37). These companies, like the military that trained many of their founders and employees, have a fascination with high technology. But, along with using the latest computers, these domestic security experts fear them as well. Computerization doesn't make us safer, it seems; rather, it makes our whole society more vulnerable. One expert claims that

> Computers are vital to everything from our national defense to our transportation networks, food distribution systems, electrical grids, and nearly every other facet of our infrastructure, and one individual who knows how the system works can inflict tremendous damage on it, and potentially even dismantle our entire economy. (quoted in Livingstone, 1990, p. 39)

So, one person could "dismantle our entire economy"! Then clearly more computer-equipped security companies are needed. But this threat is so extreme, normal rules of war will have to be abandoned:

> "In combat there is no such thing as a fair fight," observes one counterterrorism specialist. Anything short of draconian methods that will give the West an edge in its struggle against terrorism, therefore, "should be encouraged." In this connection, the war against terrorism has gone high tech and, in the final analysis, it may well be that technological superiority will be the most important factor in the West's ability to withstand the terrorist challenge. (Livingstone, 1990, pp. 139–140)

All of which makes for a nice circle.

Despite the explosion of private security, the real players in the terrorism game are the secret state agencies. While it is hard to tell information from disinformation, it seems that the U.S. agencies lead the world in computerization, especially the National Security Agency (NSA) and the Central Intelligence Agency (CIA). For them, drug wars are interesting and terrorism is significant, but it is the struggle for political power in Third World countries that is the focus of most LICs. That is the real game, and the campaigns against drugs and terrorism, as well as their covert operations that spread drugs and terrorism, are all part of the contest for political power in the United States and in other states.

This struggle has become a rationale for even the most high-tech weapons—those in space. Lockheed, for example, made up a video game display for weapon's fairs to show how easy, and how much fun, antimissile systems could be. Instead of attacking Russians, the enemy was a Libyan intercontinental ballistic missile (ICBM) (*Military Space* Staff, 1989c, p. 1). As we shall see in Chapter 3 when we consider SDI, Lowell Woods, the Lawrence Livermore bureaucrat and weapons promoter/scientist, also raised the "Third World threat" in seeking support for his brainchildren, the misnamed "brilliant pebbles" (tiny smart kill vehicles in space). They "would be," he claimed, "very, very, capable against a light launch for the Third World Country" (*Military Space* Staff, 1990a, p. 8). Drug wars and Third World conflicts have become a major justification for all sorts of space and other computerized weaponry, as can be seen in the report of the blue-ribbon Commission on Integrated Long-term Strategy.

But the commitment of the DoD to high technology in general, and computerization specifically, goes much deeper. Consider the Pentagon's own *Analytical Review of Low-Intensity Conflict* (U.S. Army, 1986), prepared by the Joint Low-Intensity Conflict Project, U.S. Army Training and Doctrine Command (JLIC).[25]

From the *Analytical Review* it is immediately clear that high technology, in general, and computers (even as metaphors), in particular, are central to all types of conflict. In official discussions computer terms ("data," "program," "access," "information") have become common. Equally prevalent is a focus on technology (logistics, weapons) and on the human as information processor (command, control, communications, psychological operations, public relations, training). One of the very first things the JLIC did was computerize itself. In terms of technology the report is also quite explicit that technology was a crucial force multiplier (p. 11-8).

Besides technology, "data" constitutes a crucial resource and "information conscious[ness]" is a central virtue (p. 7-11). "Intelligence is the most powerful tool," the JLIC claims. They go on to define it as "a unique commodity for which there is a constant requirement" (p. 12-1). Intelligence is "the means to access and influence key government policy makers" and other elites, especially the military. And it is cheap! (p. 12-1).

Information is ultimately linked with the other three elements of the old C^3I formula: command, control, communications, and intelligence. Despite the incredible communications resources of the United States, the report alleges that one of the central structural problems with command and control is that there are not enough existing "communications resources" (p. 8-2).

Sweeping claims are made for the importance of logistics. It is even argued that better logistics could have won the Vietnam War (p. 13-1). A remarkable claim in light of the thousands of billions of dollars of equipment used and squandered in that conflict. The study advocates technology that is "appropriate to the environment" and "user-friendly." A special task force, the Minor and Unconventional Warfare Project of the Army's Material

Command, was set up to identify the equipment requirements of LIC (p. 13-3). A number of pages are taken up with explaining how vital and wonderful the Army Logistics Center's logistics computer model ("Foraging") is for crucial "logistics intelligence" and coordination. Its deeper integration into training, contingency planning, and all other logistical functions is strongly advocated (p. 13-4).

Finally, there is the psychological element. "Low-intensity conflict is basically a struggle for people's minds" (p. 14-1). "And in such a battle, psychological operations are more important than fire power. . . . Insurgences, therefore, are primarily political and psychological struggles; military considerations are secondary" (p. 14-2). Psychological operations (PSYOPs) "have to be centralized and combined with all information efforts" (p. 14-5). "And the scientific exploration of PSYOP must be pursued" (p. 14-7). PSYOPs are just part of information management: "Public information, public diplomacy, psychological operations, and psychological warfare can be viewed as individual parts of a continuum of information" (p. 15-2). The report goes so far as to complain that "military public affairs personnel are extremely reluctant to associate with the PSYOP community," which is necessary in order to "bridge the gap between information and persuasion." To be successful in LIC "that bridge must be crossed" (p. 15-3). In other words controlling information is controlling people. This is especially important in respect to the military's media policy.

Another example of the central role high technology plays in low-intensity policy is this analysis of the importance of high-tech matériel for improving the chances of terrorists:

> The availability of technologically advanced weapons and communications equipment has increased the lethality, mobility, and security of terrorist and insurgent groups. Rapidly changing technology has benefited insurgents and terrorists. (p. 2-3)

Pro forma warnings of the dangers of relying too much on complex technology are included (p. 14-17), but this is not the working policy at all. Just as in Vietnam, while there is an official theoretical realization that technology can't win these types of war, the actual military practice ends up one of high technology war anyway.

In cases where there is no need for LIC theory, as with the Gulf War, the military's technophiles can go whole hog. If the enemy is incompetent, this can be incredibly effective.

Chapter Two

Computers at War:
Kuwait 1991

Large computers are still running the overall show.
—Business Week *Staff (1991c, p. 42)*

I couldn't have done it without the computers.
—*Attributed to Gen. H. Norman Schwarzkopf on TV*

There is no war, then, without representation, no
sophisticated weaponry without psychological
mystification. Weapons are tools not just of destruction but
also of perception.
—*Paul Virilio (1990, p. 6)*

We give special thanks to Mr. Bush and all the allies: the
British, the French, the Egyptians, CNN.
—*Kuwaiti man (San Jose Mercury News,*
February 28, 1991)

TVs over Baghdad

During the Kuwait war[1] the most amazing pictures were on television:
bomb's-eye views of trucks, bunkers, bridges, and runways were shown again
and again. But other images captured by this same technology—videos of
scared men running from machines and dying by machines—were censored.
One reporter who managed to see some forbidden gun camera film from
Apache helicopter raids said the Iraqi soldiers were

> like ghostly sheep flushed from a pen—bewildered and terrified, jarred
> from sleep and fleeing their bunkers under a hellish fire. One by one, they
> were cut down by attackers they could not see . . . blown to bits by bursts
> of 30mm exploding cannon shells.[2] (Balzar, 1991)

The commander of Desert Shield and Desert Storm,[3] H. Norman Schwarzkopf, called it "technology war" (*Business Week* Staff, 1991a, p. 38). Certainly for the U.S. forces it was often a technology-mediated experience with little danger for most. According to one incredible statistic it was more dangerous to be a young man back in the United States, with its automobile accidents and urban murders, than to be serving in Desert Storm! According to Prof. Charles Lave of the University of California, Berkeley, almost 300 U.S. soldiers had their lives *saved* by their service in Desert Shield and Storm (Garchik, 1991, p. A8). For Iraqis it was quite a different matter: hundreds of thousands dead and the killing continued as civil war into the spring and summer.[4] It was a war notable for its paradoxes and its intense *mediation*, and so it makes a fine case study of postmodern war.

Kuwait as a Postmodern War

Politics Is "War by Other Means"

Michel Foucault argued that today politics is war "continued by other means" in a system that is "the continuation of war" with an "end result [that] can only be . . . war" (1977, p. 244). Both President George Bush and President Saddam Hussein saw politics as war by other means during their conflict. Bush performed prodigious political feats simply to get war approved by the UN and the U.S. Congress and to prevent peace from breaking out. Saddam's foreign policy (toward Iran, Kuwait, and Israel) and much of his domestic policy (especially toward the Kurds and Shiites) is always war or threats of war. Saddam's invasion of Kuwait led to Bush's war, which in turn resulted in Iraqi civil wars. Wars breed wars. The chain goes even further back. The United States built up Iran as a subimperial power, but it became a fearful enemy after the Iranian revolution. So the United States had to build up Iraq and support its attack on Iran to balance the region. Then Iraq became the fearful enemy. Policy breeds wars.

There are other implications to the politics–war symbiosis. Many wars since 1945 have been neocolonial. The Gulf War seems like the colonial wars of the nineteenth century where a few hundred British troops with modern ships and machine guns killed tens of thousands of natives in order to make their land a protectorate or to clear them from valuable real estate. So is it a surprise that President Bush, with a move that could have been made by the British when they played the Great Game to preserve that region's balance of power, prevented the destruction of Saddam's regime, which just days before Bush had been equating to that of Hitler (Gordon and Trainor, 1995)?

No doubt Bush would have preferred it if Saddam had been overthrown and replaced by someone else. But the important considerations were that

(1) Iraq remain strong as a counterweight to Iran and Syria; (2) the Kurds not be allowed independence because that would threaten Turkey with Kurdish rebellion; and (3) a truly democratic regime not come into power that might focus its energies away from a war policy onto something more threatening to U.S. interests in the long run—a restrictive oil policy for the Middle East, for example. No doubt domestic considerations also played a role in Bush's decisions, but there the real surprise was that the great victory wasn't enough to save his presidency, even though many experts have claimed that a good little war guarantees reelection.

Any examination of the actual battle for Kuwait shows that the 1st Cavalry and the 24th Mechanized Division were kept from closing the door on the bulk of the Iraqi Army's elite Republican Guard, although the 24th did destroy 30% of the Hammurabi Division after the cease-fire.[5] The whole Allied strategy, designed as it was for retaking Kuwait, was for a limited war. Saddam's policy was also a limited one. He carefully preserved most of his Republican Guard and all of his helicopters and special forces, the best units for domestic control (Eldridge, 1991, pp. 21, 36–37; Friedman, 1991, pp. 250–260).

This war shows that winning the peace may be harder than winning the war. During the war, anti-U.S. sentiment exploded into demonstrations that forced U.S. citizens to flee Pakistan, Bangladesh, and other Moslem countries (Simmons, 1991). Months after the liberation of Kuwait, U.S. troops were ordered to invade northern Iraq to protect Kurdish refugees while other troops were still occupying the south of the country. Five years after the war Saddam was still in power, Kuwait was being ruled as autocratically as ever, and (except for the incredible pollution in the Gulf) things were much the same as they had been before the Iraqi invasion.

Informational War

Information is now the crucial military resource and information processing a central military operation. Computers were a primary weapon of the U.S.–Allied victory:

- Computers to organize and track the movement of the massive armies to the Saudi desert
- Computers to soak up and sort out the thousands of satellite images and the hours of captured electronic transmissions from the fleet of over 50 satellites
- Computers to help fly the planes, drones, and helicopters that, along with all the other weapons, were produced in computerized factories by robots as often as humans
- Computers to guide the bombs and missiles and even the artillery shells
- Computers to jam the radars and fool the targeting computers of the Iraqis

- Computers to send messages and point satellite dishes as much for CNN as for the DoD
- Computers to game the Iraqi responses and predict the region's weather
- Computers to track the weapon platforms and their maintenance.
- Computers to count the weapons expended
- Computers to look up the home addresses of the dead

Humans are still crucial for interpreting information, although there are many plans in the works to use machines in this role as well. There was also a great deal of intelligence gathering by Special Operations troops, but they too were dependent on high-tech equipment ranging from "stealth" parachutes to portable satellite links (Healy, 1991).

These thousands of computers, from ruggedized laptops to airborne mainframes, were organized into at least nine metasystems: (1) The Worldwide Military Command and Control System (WWMCCS); (2) the Modern Age Planning Program (MAPP); (3) the Joint Deployment System; (4) the Stock Control and Distribution Program; (5) the Tactical Army Combat Service Support Computer Systems (TACCS); (6) the Military Airlift Command Information Processing System (MACIPS); (7) the Airborne Battlefield Command Control Center (ABCCCIII); (8) the Joint Surveillance and Target Attack Radar System (JStars); and (9) the Operation Desert Storm Network (ODS NET).[6] A few of these networks were for watching the enemy (JStars) or simulating battle scenarios (MAPP) in Schwarzkopf's Central Command in Riyadh. But most of them were for command, control, and communications.

The ODS NET was the only one of these created especially for Desert Storm. It was managed from Fort Huachuca in Arizona and linked thousands of computers through satellites and lines over almost 50,000 miles of connections. Although it only carried unclassified traffic it was a crucial part of the operation, linking all the services at bases around the world for real-time communication. When the system overloaded on the first night of the land campaign, the system manager, Chief Warrant Officer Robert F. Weissert, was called out of bed at 2 A.M. Arizona time. Rubbing the sleep from his eyes, he went into his den and logged on to the system. By rerouting information around some overloaded computers he got the data moving within half an hour. This postmodern warrior–computer operator who battles from his den says of ODS NET: "It far surpasses anything that's ever been done in any war we've fought" (Green and Greve, 1991, p. A19). He may be right—for while topologically smaller than any other military system, it is both more flexible and more powerful then the rest. It represents another stage in the growing importance of computers to the military.

Military theorists declared years ago that for modern war you needed C^2: command and control. Early in postmodern war that became C^3I:

command, control, communications, and intelligence. In the late 1980s it became C^4I^2: command, control, communications, *computers*, intelligence, and *interoperability*. Interoperability means computers talking clearly to other computers and to humans. In this mantra, C^4I^2 computers alone among the categories are artifacts instead of ideals. Or aren't they really both?

The Iraqis responded to the computer-controlled barrage of computer-guided weapons and televised bombings with fiber-optic communication channels, numerous simulated targets, and a few stories on CNN. It was an uneven contest. High technology can be defeated, but it takes organization, loyalty, patience, political acumen, intelligence (in both senses), popular support, and allies. Saddam had few of these.

For its part, while the peace movement around the world could mobilize more antiwar protesters than could prowar advocates, it could not garner the same level of coverage that either of the warring sides could. Still, it was close. Peace seemed always about to break out. The careful deployment of images, arguments, and weapons by both Saddam and Bush kept the war alive.

So, obviously, the centrality of information to postmodern war is not limited to the battle space. Controlling information at home is just as important, and it has been Pentagon policy since Vietnam to do just that.

The Expansion of War

War has spread not just to every corner of the globe but to the very heavens as well. Gen. Merrill A. McPeak, the U.S. Air Force Chief of Staff, called it "the first war of the space age." Outer space was a crucial resource, especially for collecting information (Trux, 1991, p. 30). The war also extended across the surface of the globe. The United States showed itself quite capable of fighting a major war in the desert, which included bombing runs flown from snowy England, balmy Louisiana, and tropical Diego Garcia. The battlefield doctrine used in Kuwait was AirLand Battle, originally prepared for the European Theater. This doctrine holds that the air (including space) and the ground are equally important. Battle is three dimensional now and as it spreads into physical space it compresses in real (i.e., lived, human) time. AirLand Battle goes 24 hours a day, thanks to caffeine, amphetamines, computers, radar, and infrared. Rapid attacks, slashing and thrusting, shatter the enemy fronts and armies in a matter of hours instead of the days that blitzkriegs took, or the weeks and months of most modern-war offensives. As Virilio and Lotringer have emphasized in their work, speed becomes crucial: "There is a struggle . . . between metabolic speed, the speed of the living, and technological speed, the speed of death" (1983, p. 140). The milliseconds a laser takes to target a bunker or a jammer to confuse a radar are the margins of victory. In the war of mechanical speed against human reactions, bodies are the only real losers.

The home front is more important than ever before. The media must be guided to keep support high. Col. Darryl Henderson (USMC Ret.) notes how Gulf War bad news was managed: "restricting access to it," "presenting it as an isolated incident," and allowing it to "dribble out in a controlled seepage over a number of days or weeks." Good news is "heavily marketed," or even manufactured, to minimize bad news (1991). Along with direct news management other issues have to be dealt with, especially protests and economic strains caused by the war. Protests are managed through the media and with repression, calibrated to keep resistance low. Finally, the economy must be buffeted from the worst shocks (Reuters Staff, 1991).

The Limitations of War

Despite much bluster to the contrary, this was a very limited war. The targets open to air attacks were severely constrained for various reasons, including the fear of damaging holy sites, the fear of killing civilians openly, and a general regard for public opinion. This is not to say the bombing wasn't ferocious as hell, it was. But it could have been generally unrestrained, instead of just specifically. As free-fire zones were set up in Vietnam, "killing boxes" were established in the Desert Theater of Operations where unlimited air attacks were allowed. "Each 'killing box' is several miles long and wide and it is identified by letters and numbers, U.S. Air Force officials said" (Fulghum, 1991, p. 62). This is Cartesian war in its purest form yet.

Certain weapons, CBN (biological, chemical, nuclear), for example, were not invoked, although several U.S. generals did call for the use of nuclear bombs (Gellman, 1994). An important limiting factor was President Bush's claim that this was a "just war." Since just war doctrine calls for minimizing civilian deaths and restraint in the use of force, the open bombing of cities, as practiced in World War II, was precluded. The point wasn't so much to keep the number of dead Iraqis to a minimum. The long blockade, followed by the destruction of the country's infrastructure, including water, power, and sanitation, led to tens of thousands of civilian deaths, as did the civil war Bush encouraged. The strategy was to make it seem as if civilians were being spared. It has been at least a moderate success (Green and Greve, 1991). Even without considering that most of the Iraqi troops slaughtered by air were untrained conscripts, it is very likely that many more civilians died in this war than military personnel, continuing a trend that started in World War II.

Simulations and Games

Before it invaded Iran, the Iraqi military bought a game about such an invasion for several hundred thousand dollars from the BDM Corp., one of the top computer consultants for the Pentagon. Before the Kuwait war it brought another such game. U.S. interventions in the Middle East have been gamed and simulated for years. A specific plan for liberating Kuwait was practiced in Schwarzkopf's Central Command two years before the Iraqi invasion (Der Derian, 1991).

During the war, gaming and simulations continued. They ranged from giant mock fortifications built by the Marines in California to practice assaulting, through flight simulators used by all the pilots, to the computers at Desert Storm headquarters that ran complicated scenarios.

Humans and Machines

The U.S. military has been striving for years to replace soldiers in battle with machines so as to make foreign wars more palatable to the American people, who still refuse to admit to themselves that they are citizens of the history's most powerful imperial state. In this particular war, thanks to overwhelming air superiority and incredible Iraqi inferiority, a remarkably low casualty count was achieved, making the American public very happy. Reagan got more troops blown up in one hour in Lebanon than Bush lost in the whole Gulf War, although hundreds of thousands of Iraqi and Kuwaiti soldiers and civilians were killed. The second benefit of the machines-for-men policy is that machines help soldiers kill more enemies, not just physically but psychologically as well. It is hard to kill people hand to hand, one to one, face to face. With machines you can kill many and at a distance. This somehow seems more moral as well. The confirmed killing of thousands of civilians in Iraq would constitute a terrible war crime if they had been killed by soldiers with knives or hand guns. The real morality of the incredible air attacks can be inferred from this pilot's account of his experience: "It's almost like you flipped on the light in the kitchen at night and the cockroaches start scurrying, and we're killing them" (Morse, 1991, p. A3).

Iraqi casualties, military and civilian, were high to keep U.S. casualties low. Similar decisions have been made, albeit shamefully, during modern wars. But to bomb civilians so as to lose fewer soldiers would have been considered quite unmanly by earlier standards.

War Is No Longer So "Manly"

The foregoing makes the point that war is no longer just for men. Not that this war was less sexist than others, but it was more complicated in gender terms as the media went out of its way to show. Children were left at home with fathers when their mothers went off to war. Children were left at home parentless when their parents went off to war. Soldiers got pregnant and were sent home; others cried on TV while showing pictures of their kids to the world. But all was not gentle sweetness and light. The inevitable women prisoners and women casualties were still shocking. And on the front lines, where the living meet the dead, war's old gender rules still applied, at least symbolically.

Clearly, in war's coding, the inferior and hated enemy is feminine. Consider what one U.S. pilot said when he shot down an Iraqi: "Cold, cold smoked the bitch." The pilot didn't say "bastard," he said "bitch." In a related dynamic, covered Saudi woman were dehumanized by the U.S. troops who called them "BMOs" (black moving objects) and "Ninja women" (*Newsweek* Staff, 1991a, p. 12). And, as with most wars, while the enemy is labeled female, our weapons can be considered male. The owner of the New England Patriots football team made this clear when he compared Patriot missiles to the male genitalia his players used to harass a woman reporter: "What do the Iraqis have in common with Lisa Olson? They've both seen Patriot missiles up close" (*Newsweek* Staff, 1991b, p. 23).

Cynthia Enloe (1983, 1993) has shown that while women have always been important in war, and war has always severely impacted their lives, seldom have they participated directly in combat . Lately, however, woman have been allowed, physically and rhetorically, closer and closer to the very heart of war, the killing and dying. This also bleeds over into the militarization of the home front. During the Gulf War Tina Kerbrat, a Los Angeles policewoman, was killed by an illegal Salvadoran immigrant. Said Police Chief Darryl Gates in a "profanity-laced news conference" after the shooting:

> There's been a lot of talk about women in combat these days. . . . The Los Angeles Police Department's women are in combat all the time. There's a war right here and it's been fierce. . . . If you think the war is just in the Persian Gulf, you are wrong. Our casualties are greater in proportion to the casualties in the Persian Gulf. (Associated Press Service, 1991b, p. A6)

Judging by Chief Gates's rhetoric, by the large numbers of women in the U.S. military, and by how well they have performed, it seems clear that women are at war now to stay.

War against Nature

The Persian Gulf War was one of the most disastrous ever for the environment, especially considering the war's limited scope and duration. T. M. Hawley concludes his study of its environmental destructiveness by calculating that at one point 15% of the world's oil consumption was burning in Kuwait, producing a petroleum fog that spread over 1.3 million square miles. The oil spill was two to three times bigger than any other in history (Hawley, 1992, p. 183). It wasn't just the direct air, water, and land pollution from the combat or other hostile actions. The indirect ecological costs of supplying and maintaining the two armies, from manufacture to deployment, was at least as great as the direct battle destruction, and probably equaled whatever conscious ecological destruction Saddam ordered. Humans can totally destroy nature but the military isn't really aware of it. Bureaucratic/logistical and technoscientific power can now overwhelm the biosphere even by "accident" or "mistake" as the case may be.

War has always been polluting, and this one was no exception. In August 1990 the White House waived for the military the provisions of the National Environmental Policy Act for environmental studies because of the Persian Gulf mobilization. In January 1991 the Pentagon said that that waiver might "be the first in a wider program seeking suspension of other federal environmental statutes" (Siegel, 1991, p. B11). Ironically, just as the war was starting the DoD was committing itself to becoming ecologically responsible. But war itself is not ecologically responsible. As a Colonel Norris of the U.S. Army said in responding to a question about military pollution in Virginia, "Ma'am, we're in the business of guarding this country, not protecting the environment" (personal communication from an observer).

War as Spectacle

War was once as much a ritual as a pragmatic activity. Hunting, manslaying, and human sacrifice, the parents of war, were originally profoundly religious activities. As the child of these practices, war had as part of its very rationale a focus that linked effective and holy display with material and metaphysical efficacy. As war became civilized the show was less for fate, for the gods and goddesses, and more for the morale of the warriors on both sides. Still, Paul Virilio is right when he says, "War can never break free from the magical spectacle because its very purpose is to *produce* that spectacle" (Virilio, 1990, p. 5; emphasis in original). Without the spectacle it would not be bearable. With the Gulf War, however, a mass media war, it was not just the United States and Allied troops who were ennobled by the great display of Western weapons, it was all of the West. It wasn't just the Iraqi troops who were bedazzled by these weapons' glamorous and destructive power, although they felt it most truly, in their bodies. But it was everyone in the TV-watching world that witnessed the raids on Baghdad, that rode smart bombs into shelters and bunkers, that saw Patriots meet Scuds in cloudy night skies. We were dazzled as well, with horror, or awe, or even pleasure, or all of these at once.

And it was to us, the viewing public represented by the camera crews of the infotainment industry, that many of the Iraqi troops surrendered, sometimes even chanting "George Bush, George Bush." The line between spectacle and battle can become quite indistinct. The Pentagon contracted with National Football League films to make the DoD's official version of the war. Said NFL Films president Steve Sabol,

I don't want to say that war is the same as football, [but the military] likes the way we have presented and mythologized pro football. The same spirit and ideology that football glorifies and inspires—discipline, devotion, commitment to a cause, unselfishness, leadership—is also the spirit necessary for a successful military endeavor. (*Santa Rosa Press Democrat*, 1991, p. A1)

Meanwhile, the motion picture industry launched an offensive of Gulf War films. They included: *Desert Shield* starring Jan-Michael Vincent as a commando in the war, for 21st Century Productions; *Human Shield*, about "the kidnapping of an American colonel's brother by a vengeful Iraqi officer," for Cannon Pictures; *Shield of Honor*, an Iraqi plot to destroy Israel, starring David Carradine, and Omar Sharif as Saddam Hussein; and *Target U.S.A.*, an attempt by Iraqi terrorists to take over an American town (Arar, 1991).

J. Glenn Gray points out in his remarkable book *The Warriors: Reflections on Men in Battle* (1959) that the beauty of explosions and the eroticism of the speed and the sounds of war are very attractive to many soldiers. Now, from our living rooms, we can watch these lethal pyrotechnics in real time and with no danger, perhaps mistaking them for NFL highlights or the latest war thriller. Robert E. Lee said, "It is good that war is so horrible; lest we grow too fond of it." Postmodern war isn't horrible at all . . . for most of us.

The New Emotions of War

War has always been emotional: fearful, awful, inspiring, dreadful. Great currents of love (for comrades and distant families), pity (for enemy soldiers or civilian casualties), and hate (for the rear echelons, the brass, and civilians at home in general) have been felt by most combatants in modern war. Now, however, these emotions traditionally centered around other humans are being supplemented by emotions about machines, especially love for them (technophilia), fear of them (technophobia), and the belief that they can save us (technism). The technoeuphoria of the Gulf War makes it seem as if the high-tech weapons won the war, even though all they did was keep Allied casualties low.

Low-tech weapons worked just as well as high, and many high-tech systems had very poor performance rates or failed to work at all. The Tomahawk missiles, not used in the last half of the war, and the Patriots were particularly ineffective, although the press was told the opposite (Friedrich, 1991, pp. 169, 175). Gregg Easterbrook (1991, p. 49) concludes:

- "The low tech stuff is working as well as the high tech."
- "Obvious is doing just as well as stealth."
- "The same trends making weapons more accurate also make targets more elusive."

- "Wonder weapons kill things better than people."
- "The high tech arsenal might not be as effective against a more formidable foe."

The Patriot–Scud confrontation is a good case in point. Israeli defense officials have testified that while Patriots hit 41 of 42 Scuds that were fired at over Israel, they only destroyed the warhead 44% of the time and "may have caused more actual damage than [they] prevented" (*San Francisco Chronicle* Staff, 1991, p. A24). The 29 U.S. soldiers killed in the Saudi barracks at Riyadh may have been killed because a Patriot hit a Scud and brought the warhead down prematurely. It was headed out into the desert (Bowman, 1991, p. 8).

Finally, the very technology that brings so much speed to war also causes some surprising delays. General Schwarzkopf admitted to reporters that he couldn't change his battle plan easily because it was so computerized (Glass, 1991). In this war that didn't matter; in another war it well might.

The Special Discourse of Postmodern War

War is a discourse system, but each type of war has different rules of discourse. In postmodern war, the central role of human bodies in war is being eclipsed rhetorically by the growing importance of machines in general and weapons in particular. The number of Iraqi tanks and planes destroyed was always available; the count of dead Iraqi bodies was not. It was considered a "distasteful" or "pornographic" interest by one British briefing officer. General Schwarzkopf could even go so far as to say, "we are not in the business of killing." In his essay on bodies in the Gulf War, Hugh Gusterson (1991) explains how the general could make such a claim:

> Schwarzkopf was able to make the extraordinary and, on the face of it, absurd contention that he was fighting a war without being in the business of killing largely because of the power of a system of representations which marginalizes the presence of the body in war, fetishizes machines, and personalizes international conflicts while depersonalizing the people who die in them. (p. 51)

That war is a discourse has been remarked upon by many. The historian Susan Mansfield argued that "Traditional cultures clearly understood war as a form of discourse between the human and 'the other' " (1982, p. 236). The meticulous Quincy Wright concluded, "modern war tends to be about words more than about things, about potentialities, hopes, and aspirations more than about facts, grievances, and conditions" (1964, pp. 356–357). So it shouldn't surprise us that Schwarzkopf remarked in his famous postwar briefing, "It's hard to read an army," or that George Bush, in defending the war in his televised announcement of the hostilities, characterized it as an act of definition: "We're called upon to define who we are and what we believe."

But in postmodern war discourse the best arguments are weapons, and they trump bodies, even words, again and again. While in earlier wars the discourse was limited to those who were there, now it is a conversation for the whole electronic culture. In that conversation, images and simulations are sometimes just as important as actual events because they become events in and of themselves.

Lessons of Desert Shield/Sword/Storm

The elements of postmodern war are contradictory; so it is with the lessons of the Gulf War. War is limited in some ways, expanding in others. It is more lethal for some, less lethal for many. It remains real war in most ways, but it has become unreal in some important respects. What this war clearly showed is that contemporary war is very confusing. But three of the clearest lessons were in what the United States/UN versus Iraq war *did not* prove:

1. It didn't show that high technology and smart weapons are crucial to winning wars. The Allies would have won this war with dumb weapons. Almost all the weapons, from the dumbest to the smartest, worked with the limited efficiency of everything in war. The Iraqis had some of the newest equipment in the world. It didn't help them. This war shows how human will, accurate judgments, and political alliances are more important than ever in war today.

2. It didn't prove an end to the Vietnam syndrome. The euphoria that declared the Vietnam syndrome over with the crushing of Saddam seemed hollow less than a hundred days after headlines screamed Bush's claim that "This War Is Behind Us!" More sober analysts, including General Schwarzkopf himself, have always refused to equate the Iraqis with the Vietnamese, or the Vietnam War with shorter conflicts. To quote Schwarzkopf:

> I certainly don't give the Iraqis much credit. Ho Chi Minh and [Gen. Vo Nguyen] didn't live in luxury, didn't have seven different palaces, didn't drive white Mercedeses like Saddam Hussein. Hanoi had an entirely different class of leadership. (quoted in Lamb, 1991, p. A14)

James Webb, the marine combat veteran and former Secretary of the Navy, also refuses to claim that the easy victory in Kuwait erases the lessons of Vietnam. He points out that we didn't lose in Vietnam because of our technology but because of our opponent:

> We had one of the best-trained and best-equipped armies in American history in Vietnam. Our technology was just as good as it was in the Persian Gulf war. Not to denigrate what we accomplished against Hussein, but Hussein was no military strategist.
> If Ho Chi Minh had put 60 percent of his army in one spot where there were not any trees, we would have blown them away in 40 days too. (quoted in Capuzzo, 1991, p. E12)

That the Gulf War erases the Vietnam syndrome is just wishful and dangerous thinking. The Vietnamese won the first postmodern war because they had the proper strategy, motivation, and organization (including important allies) to match their enemies. The Soviets were driven from Afghanistan for similar reasons. Fundamentally, both were because the people of the West and North aren't willing to support long bloody wars for the sake of empire; while many people of the East and South are willing to fight and die for the hope of something better than what they have now, or at least to drive foreign armies out of their countries.

The syndrome isn't cured at home, either. There was a truly significant level of resistance to the war with Iraq, despite the brilliant political mobilization orchestrated by the war movement. Antiwar forces didn't stop the war, but they came close. The war movement did triumph in the end, but it has a problem: they may not get an enemy like Saddam again for quite some time.

3. Many analysts have been struck by the power of this war's images. These are important, especially in justifying war and seducing many to its glamour. But simulations are not as important as the real, Baudrillard (1991) to the contrary. For those who were there, or anyone who reflects seriously on what being there meant, this was war like war has been for thousands of years—death, fear, relief, and many more emotions, and real bodies really dead at another's hand, no matter how remote. This is the real horror, and it holds the true glamour in its old twisted rituals.

The New World Disorder

> This is black versus white, good versus evil.
> —*President George Bush, to a group of*
> *college students in January 1991*

> Two distinct, through interrelated, forces have shaped modern Western culture and identity. The first . . . has been the centrality of reason as the constitutive principle of modernity itself. . . . The second . . . has been the intensive

and intense encounter with other cultures brought about by
imperialism expansion. . . . Saddam is not an alien monster, a
monster against modernity, but rather a monster born of
modernity, a monster within modernity.
—*Kevin Robins (1991, p. 42)*

The New World Order—Peace in the West, war for the rest.
—*John Brown Childs (1991, p. 83)*

Horrible as the Persian Gulf war was, it didn't mark the beginning of any
new era or new type of war. The New World Order is the Old World
Order. The United States had made a commitment to the continuation
of the Cold War militarization of domestic and foreign policy even as
the Soviet empire began its collapse. As Marine Col. Lawrence Karch
put it in "The Corps in 2001":

> Increasing nationalism, religious and racial strife, bitter sectarian enmities,
> competition for materials and energy, and endemic poverty will ensure a
> slow burn into the 21st century. . . . The United States for its part must be
> a steadfast "Arsenal of Democracy" in low-intensity warfare, and a com-
> batant only when U.S. national interests cannot otherwise be protected.
> (1988, p. 41)

He went on to predict numerous LICs and at least one midintensity war
involving the United States, most likely in Korea, Southwest Asia, or
the Mediterranean. Sociologist John Brown Childs quotes an article by
Gen. George Crist (also of the Marines although retired) in the winter
1990 issue of *Strategic Review*. General Crist makes it clear that as the
Soviet threat decreases "operations in the third world move up in
priority" (Childs, 1991, p. 6). It is called by some the "Rogue State
Doctrine," and it has justified a U.S. military in the mid-1990s of roughly
1.5 million soldiers and a yearly war budget under President Clinton of
$260–270 billion, which is equivalent to what the United States spent
yearly during the Cold War, except during the Korean and Vietnam Wars
(Klare, 1995, p. 625).

Noam Chomsky (1989) quotes Dimitri Simes, a well-connected academic, regarding one version of this new "American Peace" that many
governmental voices have called for. It doesn't seem that peaceful at all:

> First, the U.S. can shift NATO costs to its European competitors. . . .
> Second, it can end the manipulation of America by Third World nations.
> . . . Third . . . the apparent decline of the Soviet threat . . . makes military
> power more useful as a United States foreign policy instrument. (p. 7)

To justify this new aggressiveness, the military has to claim that today's world
is *even more* dangerous than the superpower standoff. By 1991 this had

become the U.S. Army's official "Posture Statement." "The United States faces as a complex and varied a security environment as it enters the 1990s as in any time in its history" (Klare, 1990).

Along with the continuation of cold (or low-intensity) war in its new diffuse North–South format, there are a number of aspects of the world situation that are important for understanding postmodern war. Across the spectrum, enumerated below, a number of the usual signs of postmodernism are present, including the centrality of computation and information, the increases in speed, the proliferation of contradictory trends, strange marriages and alliances, machine mediations of culture, and so on.

*   *   *

There are those who have nothing but praise for how the United States methodically destroyed Iraq and its armies with technoscience. They see this as a cool, reasonable way to wage war. But it is actually a kind of technological fanaticism, as a brief reflection on the fate of the Iraqi conscripts shows. Michael Sherry, in his comments on the United States's strategic bombing of Japan, describes the technological fanaticism of U.S. military policy in Iraq perfectly:

> The lack of a proclaimed intent to destroy, the sense of being driven by the twin demands of bureaucracy and technology, distinguished America's technological fanaticism from its enemies' ideological fanaticism. That both were fanatical was not easily recognizable at the time because the forms were so different. The enemy . . . had little choice but to be profligate in the expenditure of manpower and therefore in the fervid exhortation of men to hatred and sacrifice—they were not, and knew they were not, a match in economic and technological terms for the Allies. The United States had different resources with which to be fanatical: resources allowing it to take the lives of others more than its own, ones whose accompanying rhetoric of technique disguised the will to destroy. As lavish with machines as the enemy was with men, Americans appeared to themselves to practice restraint, to be immune from the passion to destroy that characterized their enemies and from the urge to self-destruction as well. (1987, pp. 253–254)

How this technological fanaticism has come to dominate U.S. military policy is a long story, some of which is recounted in this book. But before we look at its history in detail we should first get a better measure of its scope. The 1990–1991 Mideast War is a good case study, but it is not unique. Technoscience is central to the rest of U.S. military policy as well. It turns out Desert Storm was just an expression, albeit the first really effective one, of a philosophy of war that is in place throughout the U.S. Department of Defense, in all branches, for almost all missions, as the next chapter shows.

Chapter Three

Military Computerdom

> Our leaders and scholars . . . have given up on peace on
> earth and now seek peace of mind through the worship of
> new techno-dieties. They look up to the surveillance
> satellite, deep into the entrails of electronic micro-circuitry,
> and from behind Stealth protection to find the omniscient
> machines and incontrovertible signs that can help us see
> and, if state reason necessitates, evade or destroy the other.
> —*James Der Derian (1990, p. 298)*

The Proliferation of Military Computing

Before the cyberwar mania, described in Chapter 1, and the computerized
Gulf War, of Chapter 2, most of what the public heard about military high
computing was about the Strategic Defense Initiative (SDI)[1] and its proposed
artificial intelligence (AI) control system. There was also discussion among
computer scientists about the related Strategic Computing Program (SCP).[2]
Both these projects will be described in detail below, but first it must be
stressed that they are only part of a massive commitment since the 1980s to
AI and other advanced computing technologies by the U.S. military. Expectations that this commitment will eventually pay off are what underlay the
theories of info- and cyberwar.

Projects using AI and related techniques include "brilliant" or "autonomous" missiles (they choose their own targets), a "robotic obstacle-breaching
assault tank" (a mine remover), an autonomous vehicle for laying mines
underwater, all sorts of automated construction and manufacturing equipment, a wide range of robotic research, numerous programs for automating
the maintenance of high-tech weapons, and an ever-expanding army of data
control and manipulation programs.

The U.S. military has a large AI research capability of its own. The Air
Force has AI institutes at Wright Aeronautical Laboratories, the Human
Resources Center in San Antonio, Logistics Command, the Institute of

Technology, and the Avionics and Materials Laboratories, as well as an expert systems focus and new funding for the Rome AFB Air Development Center. The Army has kept up, with AI centers in the Pentagon, at the labs at Aberdeen, Maryland, and several other posts. Inevitably, the Navy also has a number of AI and robotic research centers, including one in Washington, D.C. (Naval Center for Applied Research in AI) and others at the Naval Ocean Systems Center and the Naval Supply Systems Command. Finally, the Joint Chiefs use expert systems in their logistics directorate to "help plan and execute joint military operations."[3] These, and new facilities, some under the aegis of the Advanced Research Projects Agency (ARPA), will do the research for cyberwar, including the $2 billion in research on computer security and related topics approved in 1995.[4]

The national labs are also heavily involved in military computer research. At Livermore there is research on optical logics, parallel computing, and supercomputers. Sandia has chosen AI work in target discrimination for special emphasis (*Aviation Week & Space Technology* Staff, 1985, p. 36). There is SCP-related research at the Jet Propulsion Labs and at the National Institute of Standards and Technology, formerly the National Bureau of Standards (DARPA, 1987), and at the Oak Ridge National Laboratory there is work in conjunction with the U.S. Army Human Engineering Laboratory and the Air Force's Wright Aeronautical Labs and for Star Wars (Weisbin, 1987, pp. 62–74).

All in all there were over a third of a million computers in the Pentagon's inventory in 1989. Computer spending by the DoD was $14 billion in 1985, and it doubled by 1990. Digitalization alone is slated for $7.9 billion for 1995–2000.[5] It is one of the few areas exempt from the Cold War retrenchment. There has been a corresponding increase in AI research in closely related non-Pentagon military areas such as intelligence, especially the NSA and the CIA, and other investigative/enforcement agencies the: ATF (Bureau of Alcohol, Tobacco, and Firearms), IRS, DEA, Secret Service, and FBI (Schrage, 1986). *Christian Science Monitor* reporter Warren Richey found that the FBI's expert system, for example, is supposed to advise agents on "when to initiate surveillance of a suspect, when to make an arrest, and even when to drop an investigation" (1986, p. 3). The FBI is playing a major role in debates about computer security as it constantly intervenes in legislative debates on the side of increased surveillance (Warren, 1996).

Of course many other military powers are committed to computing. There are major programs in the United Kingdom, Israel, Germany, Australia (Smith and Whitelaw, 1987), China, and Russia and other parts of the former Soviet Union, among others. Soviet policy was remarkably like U.S. plans, including several battle-manager-type programs called *Konsul'tant* and *Pomoshnik*, the "assistant" (Druzhinin and Kontorov, 1972). Israel's military was using an expert system for administration in the 1980s and has made important advances in vision research (Elbit computers) among other fields.

Germany is working closely with the United States on expert systems for tactical aircraft as well as other projects. China's military AI research is mainly in the areas of simulation, translation, and training. Japan, with one of the largest and fastest growing defense budgets, probably has a large military computing program, but it is veiled in secrecy.

The Strategic Computing Program and Strategic Defense Initiative

With the SCP, formally started in October 1983, the DoD was trying to develop the next generation of computing technology within a military context (Guice, 1994). The SDI battle manager, the computer system that was supposed to control the SDI sensors and weapons, was the largest and most complex software project up to that time. It was conceptually (though not administratively) a part of SCP. Making the scientific breakthroughs in computing that the SDI needed was a key goal of the SCP.

The original SCP demonstration projects (the autonomous land vehicle, pilot's associate, and naval battle manager) were augmented by the AirLand Battle (ALB) manager, radar imagery analysis, and smart weapons, all examples of concrete military applications. There was also the related submarine commander's associate program. Flashy as these projects were, what made SCP particularly important was its attempt to revolutionize machine intelligence. Without basic breakthroughs in computer architecture, symbolic processing, and other areas, these military AI projects would be impossible to produce. As Stephen Squires, then assistant director of DARPA's Information Processing Techniques Office, put it, "We're really trying to develop a fundamentally new type of computing technology, not simply new hardware" (*Aviation Week & Space Technology* Staff, 1986, p. 49).

The demonstration projects depended on advances in a number of subfields: mechanical vision, speech recognition, natural language processing, and expert systems technology. The work in these areas was dependent, in turn, on innovations and discoveries in the areas of hardware and software development that were supposed to increase computer performances many times over, as well as produce vast improvements in automated manufacturing and technology transfer between the military and civilian sectors of the computer industry.

Most of these goals were not met. The autonomous land vehicle prototype, for example, was terminated because of such difficulties as distinguishing shadows from objects and roadways from ditches. While there were discoveries and improvements in multiple processing and speech recognition, in other areas there was minimal progress (Pollack, 1989).

For another example, the original SCP report predicted reasoning with uncertainty abilities in place for battle management by the end of 1985

(DARPA, 1983, chart W6196CJ3285). Yet, at the battle management workshop at the AAAI conference in 1987 various approaches to reasoning with uncertainty were hotly debated, with the only consensus being that any workable solution had yet to be demonstrated. This situation was unchanged in 1995, ten years past the date of the scheduled breakthrough. Similar delays can be noted in many other difficult software tasks.

Many critics expected this difference between meeting hardware and software benchmarks (C. H. Gray 1989). Hardware improvements are predictable and regular. Some observers predicted that when goals were not met in the more subjective area of software development, SCP work would go forward anyway because of the many pressures to produce and deploy something to show for the effort. Two University of California professors who have studied these projects note:

> Once vast sums of money have been spent, the temptation will be great to justify the expenditure by installing questionable AI-based technologies in a variety of critical contexts—from data reduction to battle management. (Dreyfus and Dreyfus, 1985, p. 8)

Despite the sustained and constant criticism from scientists and others, both the SCP and SDI continued to be funded at extraordinary levels by the government. The second five years of SCP funding was almost twice the $339 million appropriated for the first five years (DARPA, 1987, p. 26).

After ten years many key SDI projects continued to survive. As it became obvious to everyone that SDI's original goals were impossible to meet, a number of scaled-down proposals surfaced. An incomplete space-based defense system is supposed to have some value in shooting down a few ballistic missiles from some Third World country, if it is nice enough not to send them in a briefcase, as a car bomb, or on a cruise missile. Even as the Gulf War raged, the military was using it as a justification for continuing Star Wars research, relabeled officially GPALS (Global Protection Against Limited Strikes) (Klein, 1991). In early 1992, the Russian Government proposed the construction of a First World system, including Russia, the United States, NATO, and Japan, aimed at preventing Third World ICBM attacks—as if such a system would help against nuclear proliferation. But it would consolidate space as the high ground for the wealthiest countries and lead to its further militarization.

SDI has been more of a political proposal than a technical one ever since Edward Teller whispered it into Ronald Reagan's ear. Teller's protégé at the Lawrence Livermore Laboratories, Lowell Wood, tried to save SDI with his "brilliant pebbles" proposal, first made public as a *Reader's Digest* article. He advocated putting 4,614 tiny kinetic kill vehicles with their own artificial intelligences into orbit at 450 km to attack any ballistic missile threat. These remarkable little dream weapons were estimated to cost $1.4 million each in 1990, up from an estimate of $100,000 each in 1988. Wood claimed they would be "very capable" against a "light" attack from a Third World country. Still, according to the trade journal *Military Space*, the "technical strains over Brilliant Pebbles are matched by a search for political justification" (*Military Space* Staff, 1990a, pp. 1, 7–8).

Brilliant pebbles aside, computing was the most intractable problem of SDI. As one SDI computing researcher told *Military Space*:

> Progress has been uneven. Some areas, such as weapon target assignment algorithms, have proved to be easier than expected. In other areas, such as software producibility and man-in-the-loop, we don't even know all the questions we need to ask. (1988b, p. 1)

DARPA claimed great success for the SCP, and the industry press usually agreed. There was real progress in hardware production, increased processing speeds, and the successful prototyping of some of the new symbolic processors that calculate with logical propositions instead of numbers. In other areas, especially software development and AI production, there was failure.

The Military–Industrial Community

Insiders call it a community. From outside it has been labeled the iron triangle, the military–industrial complex, and the culture of procurement. Since the 1950s thousands of high-ranking military and civilian DoD employees have gone to work for defense contractors and thousands of contractor employees have gotten DoD jobs managing contracts (Fallows, 1982, p. 65).

Consider the case of Advanced Decision Systems (ADS), a California firm with AI contracts for Star Wars and the AirLand battle manager. *San Jose Mercury* reporter Susan Faludi (1986) discovered that a number of top ADS managers come from the same halls of the Pentagon where the contracts originated. One common connection is to have industry scientists sit on the advisory panels that recommend more of their products to the military. ADS enjoyed a number of such interfaces. It doesn't make for the most rigorous oversight. Many of the computer scientists at ADS admitted to Faludi that there was no possibility of meeting the claims in their contract bids. Said one young computer whiz of the Pentagon brass he worked with, "Most of what my clients want is fantasy. You tell them whatever they want to hear and then you do whatever you want" (p. 13).

It is the same with other programs. Senator Carl Levin's Subcommittee on Government Management found that out of ten experts reporting from the SDI organization's advisory committee on computers, five were paid consultants of concerns doing SDI-related work, six were directors or trustees of such organizations, and eight were stockholders in such firms or institutions. One man was "a paid consultant to seven SDI firms, a director of one and a stockholder in seven" (D. Johnson, 1988, p. A1).

The clearest connection of the SCP was with the established military bureaucracies and companies. Numerous military commands (16 are listed as part of the SCP in its founding [DARPA report, 1983, p. 35]) and all the services were deeply involved in the SCP and other AI applications. Outside the Pentagon industrial and academic participation was also substantial.

Most of the proliferation of AI companies and new AI divisions in established corporations during the 1980s can be traced to the military. Traditional defense contractors either set up their own AI units (TRW, Martin Marietta, Lockheed, Boeing, Rockwell International, General Electric, Goodyear, McDonnell Douglas, Texas Instruments, MITRE, Hughes, GTE, General Dynamics, and Ford Motor Co.), or bought into smaller companies as Ford (Inference Co. and Carnegie Group), Raytheon (LISP Machines), Proctor & Gamble (10 percent of Teknowledge) and General Motors (all of Hughes and 10 percent of Teknowledge) have. A glance at the SCP contractors[6] shows the extent of this development as well as the involvement of most middle-sized AI firms in military computing and the existence of many little AI companies totally dependent on military contracts.

Dependency also describes many academic AI programs.[7] It is a conscious policy of the military to shape AI research to military needs. In congressional testimony Dr. Robert Cooper, head of DARPA at the time, testified that his office planned to quadruple the number of AI graduate students supported by 1986 to at least 500. He said, "We expect to develop a whole new generation of computer scientists who are in parlance of this [military] technology and are competent artificial intelligence experts" (Congress of the United States, 1985, p. 648). A former DARPA bureaucrat, Stephen Andriole declared, "DARPA is responsible in large part for the existence of applied AI" (see Andriole and Hopple, 1989, p. xii, n. 4). Robert Sproul, a former DARPA director, echoes this: "all the ideas that are going into the fifth-generation project—artificial intelligence, parallel computing, speech understanding, natural language programming—ultimately started from DARPA-funded research" (*IEEE Spectrum* Staff, 1984, pp. 72–73). In 1984, some 60 percent of academic computer science research was funded directly through the DoD, a figure that has probably increased since (C. Thompson, 1984). Some computer scientists claim that this militarization is not beneficial to AI as a science (Weizenbaum, 1976; Winograd, 1987; Yudken and Simon, 1989). As a case in point, Georgia Tech has become so much a military lab that only students with security clearances can have full access to research resources (Sanger, 1985). This dynamic has remained unchanged into the mid-1990s. When federal budgets are cut it is not military research that gets trimmed (Cordes, 1995).

Not only is the development of industrial and academic AI entangled with military priorities, but political and military policy has become intertwined, even twisted, with illusions about computerization.

Successes and Failures

Because the United States lost the Vietnam War some people said that computerized high technology was useless. Because the United States won the Gulf War some people said that the same technology is always victorious. Neither assertion is true. A technology may be powerful, but unless it is deployed at the right time in the right way it will not be useful as a weapon; it might even do more harm than good if it leads to the wrong strategy. The problem is that the reasons for using most weapons are often not logical or utilitarian; sometimes they are emotional and political. This encourages systematic misapplications leading to mistakes. The most common erroneous judgments of high-tech weapons and logical systems involve either misestimating their effectiveness or misapplying their power.

These errors include:

1. *Overestimating the effectiveness of such weapons systems*—the most common error, as can be seen with strategic bombing in World War II, the electronic battlefield in Vietnam and other deployments by the U.S. military, and the USSR in Afghanistan
2. *Underestimating their effectiveness*—as the Arab military has consistently misjudged Israeli avionic and drone technology
3. *Not knowing when to use the technology*—clearly seen in the overinvolvement of Washington, D.C., in overseas actions because of improved satellite communications

A review of the military's experience with computerized high technology shows a number of successful cases, but also a shocking number of failures.

Robotics, Automatic Manufacturing, and AI Maintenance

Prof. Bernard Roth of Stanford University, a leading robot researcher, surveyed military robotics and concluded "it is extremely unlikely that autonomous robots using the current technology or logical near-term developments from this technology will be of any real military value" (1987, p. 6).

Roth includes the military's robotic manufacturing in his indictment. This fits with the recent history of the U.S. military's attempts to automate factories, which have left the U.S. civilian robot industry behind that of the Japanese (Noble, 1986a). There have also been severe problems with the semiautomated production of high-tech weapons such as the Harpoon, Phoenix, and Sparrow missiles (M. Thompson, 1988).

DARPA has funded three- and four-legged hoppers and runners at Carnegie–Mellon University, a six-legged machine at Ohio State, and robot hands at Stanford University and the University of Utah. The Air Force hired Marvin the Robot for minor maintenance, runway repairs, fire fighting, and rearming of aircraft while a base is under attack. The Army is developing robots for everything from security to the collection of biochemically contaminated human remains (*San Jose Mercury News* Staff, 1987). The Navy and Air Force are also developing teleoperated systems (Aronson, 1984). Actually, most of what the military calls robots are really these teleoperated systems, controlled from afar. Humans are always "in the loop."

The B-1 bomber is dependent on a number of automated repair programs including expert systems.[8] Equipment breakdowns are far above predicted levels, and the number of planes on alert will be lower than planned because of them (Moore, 1987). The SCP project for a pilot's associate is really five expert systems linked together to help the pilot fly, and even fight, the plane. The proposed AirLand battle manager, for another example, will operate at the corp level, and below, during general war. It will try and take in all the possible information and display what it has been programmed to consider relevant to the human officers. It is also supposed to eventually be able to offer advice on what the enemy might be doing and how to counter it, and even make troop deployment and logistical decisions to implement its plans, and then cut and print out the orders itself. There were over 20 military research projects on expert systems in 1988 (Shah and Buckler, 1988, pp. 15–21).

Expert systems that are to perform in domains where there is little human expertise, or where what expertness there is cannot be reduced to computer code, have a very poor record. Those two University of California professors (of philosophy and engineering respectively), the brothers Hubert and Stuart Dreyfus, argue in *Mind over Machine* (1985) that the best an AI system can be is competent, since expertness depends on intuition, experience, and other intangibles.

In the long run it does seem that limited "bit" AI systems can eventually contribute to improved maintenance of military weapons and platforms, but despite much effort coordinated by the Joint Services Working Group on Artificial Intelligence in Maintenance, little can be expected until the next century (Richardson, 1985).

Intelligence (Cryptology, Surveillance, Image and Sound Analysis, Satellite Control, Database Management)

The greatest and clearest successes of advanced computing are in this area. Computers are for handling information above all. While shrouded in secrecy, CIA and NSA code breaking is reputed to be quite successful, especially against the poorer countries and companies. Sound and image surveillance (using various spectrums, including some that can probe more than 60 feet into the ground) is also a crucial source of accurate intelligence. Current systems are totally dependent on advanced computer techniques, including sorting of target data, enhancement of sounds and images, and the control of satellites and other collection platforms. However, there have been some serious computer problems at the Consolidated Space Test Center, the Blue Cube (S. Johnson, 1988). Most importantly, U.S. intelligence analysts have failed again and again to predict major events such as the Soviet invasion of Czechoslovakia, the Tet offensive, the collapse of communism, and the Iraqi invasion of Kuwait. No matter how much information machines collect, it is useless unless it is understood.

Autonomous Weapons (Shoot 'n' Scoot, Fire 'n' Forget, Self-Contained Launch-and-Leave First-Pass Single-Shot Kills)

While truly autonomous weapons are still in the distant future, so-called smart weapons, which follow carefully programmed instructions to find targets that they then compare to stored images, did work in the Gulf War, especially the Tomahawk cruise missiles launched by the Navy. But, since they needed individual programming and carried only a limited payload, they were useful only on very short lists of targets. The British deployed autonomous weapons with limited success during the Falkland War (Military Technology Staff, 1987, p. 52). The Sea Wolf antiaircraft system in particular had many computer problems, including resetting itself when confronted with multiple targets (Hastings and Jenkins 1989).

Teleoperated (remote-controlled) systems have had a somewhat better record. Israeli drones have proved very valuable in drawing antiaircraft fire. After many expensive failures going back to Vietnam, remote piloted vehicles (RPVs) had moderate success during the Gulf War directing battleship guns and Marine Corps artillery (Frantz, 1991). Still, the U.S. high-tech drone programs, bedeviled by failures and cost overruns (M. Thompson, 1988) have never been as successful as the Israeli low-tech approach (Hellman, 1987).

The U.S. Army bought a whole "family of three types of unmanned air vehicles" even though a fly-off between two vehicles (one from California Microwave and the other from Lear Siegler's Developmental Science Corp.)

resulted in both failing. The three drones perform imagery intelligence, signals intelligence measurement and signatures analysis, and meteorological data collection; they also can operate nuclear, biological, and chemical detection sensors (*Defense Electronics* Staff, 1987, p. 37). The Gnat 750 was tested in Bosnia in 1995. The more elaborate Predator and the high-tech Dark Star were still under development in 1996, but some of their supporters were already predicting the end of manned aircraft (Patton, 1996). What will undoubtedly actually happen is that drones will be used for real-time intelligence, decoys, and artillery spotting and human warriors will still fly planes long into the twenty-first century.

Modeling, Testing, and Simulation

In 1986 Wayne Biddle revealed the details of a series of weapons tests, including the testing of the Aegis naval antiaircraft system in a corn field in the Midwest and the testing of laser-guided Paveway bombs by aiming the targeting lasers through long plastic tubes so they would not be obscured by dust. Is it any surprise that four of nine laser guided bombs failed during the attempt to kill Col. Mu'ammar Gadhafi and his family (Hersh, 1987)? Biddle sums up by noting, "Our weapons tests now use so much computer modeling and simulation that no one knows whether some new arms really work." In the case of the Divad air defense gun, target drones allegedly downed by the gun were actually destroyed by radio control from the ground. The gun was canceled after $1.8 billion had already been spent (Biddle, 1986, pp. 50–66; Grier, 1986a, b). This same problem applies to nuclear weapons (Anderson, 1984, pp. 130–137).

Since it was impossible to actually test the SDI system, it was modeled and simulated instead. Martin Marietta won a contract worth over $1 billion to build a National Test Bed to do this. More than 2,300 computer and military experts worked there in a complex that included two Cray supercomputers, an experimental IBM supercomputer, and many smaller dedicated battle management stations. *The New York Times* Service (1988) reported, "The test bed's findings are expected to be critical in reaching a decision in the early 1990s over whether to deploy a rudimentary SDI system." Not surprisingly critics argued that "the system can be manipulated to yield any results the Pentagon seeks."

A study group from Computer Professionals for Social Responsibility concluded:

> The National Test Bed is being built to answer the very important question, "Will the SDI work?" But the Test Bed cannot answer this question.... Any "answer" provided by the NTB will be a very expensive guess—a guess that

will bear the seductive authority of scientific analysis and highly classified data. (Parnas and the National Test Bed Study Group, 1988, p. 6)

The United States has also made an extensive commitment to computer-mediated war gaming. The Pentagon has a whole bureau focused on computer gaming; once called the Studies, Analysis and Gaming Agency, it is now the Joint Analysis Directorate. Military commanders throughout the world are hooked into the SINBAC (System for Integrated Nuclear Battle Analysis Calculus) computer network for preparing contingency plans. RAND Corp. won a competition for the next generation of war games with the AI-based RAND Strategy Assessment System (RSAS), which includes computer programs ("agents") to play against the humans or each other. These two high-tech approaches are only the tip of the gaming iceberg. In 1982 (the last study available) there were 363 different war games, simulations, exercises, and models in the Pentagon's arsenal.[10]

The participation of the highest U.S. officials in such gaming shows how important it is. The Chairman of the Joint Chiefs of Staff chooses four or five major games each year. High officials take part in the games themselves. One in 1982 included such luminaries as former Secretary of State William Rogers (who played president), former CIA Director Richard Helms (as vice president), and numerous active Reagan appointees like Fred Ikle (then Undersecretary of Defense) and Walter Stoessel (then Deputy Secretary of State). This particular game ended when pseudo-Vice President Helms, acting after a Soviet decapitation attack had killed pseudo-President Rogers, launched a full nuclear strike on the USSR to the applause of the other game participants. Real President Reagan called the gamers up to thank them (Allen, 1987, pp. 4–304 passim; Masterson and Tritten, 1987). It isn't only nuclear war that is simulated; everything from flying planes to shooting them down is now played and practiced on training machines.

In 1983 the military and NASA spent $1.6 billion on flight simulators, missile trainers, and other such devices. By 1990 the figure had at least doubled, and two-thirds of the total global simulation and training business was with the U.S. military. In 1993 the military's ten year plan was to spend $370 billion on electronics over the next decade, much of it in simulations (Sugawara, 1990, p. 51).

One key site is the NASA–Army lab at Ames Research Center. There a complex system simulates helicopter flying under combat conditions. The first scenario simulated was war in Europe, but the Middle East and a tropical combat experience were added. The chief of the research, Maj. James Voohees, is very proud of his machines. "The human is the limiting factor," he proclaims. As with many such simulations, because of delays in hardware and software production it is often necessary to resort to simulation of the simulators. In this case 12 Amiga personal computers were used to emulate the simulators (J. R. Wilson, 1988, pp. 1127–1129). While simulators were used extensively to prepare the soldiers who fought in the Persian Gulf, some experts complain that simulators don't take human factors into account, nor the mentality of the enemy (Sugawara, 1990).

The quest in ongoing for "synthetic environments" and "seamless simulation." Sometimes the strategy is to work backward. The U.S. Army and ARPA (Advanced Research Projects Agency) have created an incredibly accurate simulation of the Battle of 75 Easting from the Gulf War, where the 2nd Armored Cavalry Regiment fought the Tawakalna Division of the Iraqi Republican Guard. The military claims, justifiably it seems, that this battle is the most accurately recorded combat engagement in human history. Bruce Sterling calls the simulation a "complete and utter triumph of chilling analytic, cybernetic rationality over chaotic, real life human desperation." He concludes that it is "virtual reality as a new way of knowledge: a new and terrible kind of transcendent military power" (Sterling, 1993, p. 52). But as impressive an experience as it is, will it really help U.S. soldiers fight better in other kinds of battles?

The most ambitious simulation since the SDI test bed is the Air Force's project to design air combat simulators with intelligent artificial agents participating equally with humans on simulators. Despite the traditional early enthusiasm, the program seems bedeviled by problems. Yet the Chief of Staff of the U.S. Army can still claim in 1995 that the combination of live, virtual, and constructive simulations means that any option conceivable can be simulated (Sullivan, 1995, p. 10). The real problem is not simulator quality anyway, nor finding opponents to play with. The real problem is knowing what to practice for in postmodern war.

C^4I^2 (Command, Control, Communications, Computers, Intelligence, and Interoperability)

Adm. William A. Owens (1995) is right when he argues that this is now a system of systems, linking 14 major sensor systems through 14 major C^4I^2 nets to 14 precision-guided munitions. But while there have been remarkable increases in communications technology, there are still many operational failures (Gilmartin, 1987, p. 11; Marcus and Leopold, 1987; Wolfe, 1985), going back to the Worldwide Military Command and Control System (WWMCCS, pronounced "WIMEX"). Theodore Roszak reports:

> WIMEX is a global network of sensors, satellites, and computer facilities that ties together twenty-six major American command centers around the world. Since the mid-1970s, WIMEX has been the subject of frequent

and urgent criticism. . . . In 1977, during a [system-wide] test, more than 60 percent of its messages failed to get transmitted. During one eighteen-month period in the late 1970s, the North American Air Defense reported 150 "serious" false alarms, four of which resulted in B-52 bomber crews starting their engines; missile crews and submarine commanders were placed on high alert. (1986, p. 195)

When the system works it can even be worse. Consider the botched USS *Mayaguez* rescue where the Marines staged a bloody and useless attack after the ship had been released. The commander of the rescue, Maj. Gen. A. M. Gray, "became so confused by the various orders coming from Washington—which helicopters he was to use, where various ships should proceed, which bombs to drop—that he just turned his radios off" (Coates and Kilian, 1984, p. 65). Similar problems bedeviled the Iran rescue mission and the invasion of Grenada (Fallows, 1982; Coates and Kilian, 1984), and they haunt the command and control of nuclear weapons as well (Ford, 1985; Pringle and Arkin, 1983).

Because of—or despite—this record, military budgets reflect an extraordinary investment in future C^4I^2. Funding has increased from $10 billion in 1981 to $22 billion in 1987. In 1989 requests were for over $24.5 billion (Lantham, 1987, pp. 18–21). This included the new Pentagon crisis center run by a VAX 8550 mainframe computer from Digital Equipment Corp. and Apollo workstations (Marcus, 1987). The Air Force–Army Joint Surveillance Target Radar System will have over 2 million lines of program code (Editors of Time–Life, 1988, p. 98).

The nation's official emergency management headquarters, the National Security Council Crisis Control Center, dates back to 1983. It was first tried out during the Grenada operation, when it was personally managed by Col. Oliver North. The highly computerized Crisis Management Support and Planning Group within the National Security Council was first entrusted to Richard Beal, a political scientist who had worked for Reagan's pollster, Richard Wirthlin. Beal set up an elaborate graphics system so President Reagan could watch simple graphic and video presentations on world trouble spots instead of having to read about them. In 1984 the NSA bought several hundred AT&T computers for almost $1 billion to be distributed to NSA outposts around the world. They are tied into the crisis management headquarters in Washington, D.C. (Roszak, 1986, pp. 201–202).

The CIA continually upgrades its AI brain, SAFE, which computer-sorts messages from around the world (Beers, 1987). Ornate command and control systems remain the military's ideal despite the extreme problems that arise in making them work, even in peacetime (Baranauskas, 1987; Gordon, 1987; Nikutta, 1987; Seiffert, 1987).

Battle Managers

Many objections to AI battle managers have been raised. They point out that the complex decisions command in war involves are difficult, if not impossible, to cast into a set of rules and calculations (Chapman, 1987b; Snow, 1987). Even more dubious is the top-secret project to build the battle management system that would fight World War IV. Yes, World War IV. Hundreds of millions of dollars have been spent to develop the Milstar and Navstar satellites and the Looking Glass command planes to fight another nuclear war after fighting a six-month-long nuclear World War III (Weiner, 1990).

What these problems might mean in detail can be seen by a case study of the most sophisticated weapon system yet deployed, the Aegis.

Aegis: Artificial Intelligence at War

Well, to me Aegis is an expert system in many respects. In fact, it has overcome much of the propensity to make mistakes through expert system instrumentation.
—Rear Adm. Wayne Meyer, Aegis manager 1970–1983[11]

Aegis is truly Star Wars at Sea.
—An unnamed U.S. Navy Admiral (quoted by Congressman Charles E. Bennett [D-FL], Congress of the United States, 1988, p. 16)

During operations against Libya in the spring of 1986, the USS Yorktown's Aegis antiaircraft system (SPY-1 phased array radar, Standard 2 guided missiles, and Phalanx machine guns linked by computer so that Aegis can act under human control or autonomously), identified an unknown target on its own. At the captain's order two Harpoon ship-to-ship missiles were fired at the unidentified radar contact, which then disappeared. At first it was claimed that the ship had sunk a Libyan patrol boat, but now the Navy admits that perhaps it was just a low-flying cloud (Grier, 1986a, p. 3). It could have been some fishermen or pleasure boaters as well who were harpooned. Nobody will ever be sure.

Another incident worth considering is the Iraqi Exocet missile attack on the USS Stark in May 1987. While the Stark was not defended by an Aegis system, certain key Aegis subcomponents were on board. It is significant that the Phalanx automatic gun system (20-millimeter machine guns that can shoot 3,000 rounds a minute with built-in targeting radar and autonomous firing abilities if put on automatic) failed to detect the incoming missiles, as did the Stark's Mark 92 fire-control radar and the SLQ 032 radar sensor. There was no chance of the Phalanx shooting down the Exocet missile because the Phalanx system was not on automatic, but it should have picked up the missile on its targeting radar and given off a loud alarm. Some analysts claim it did not because the missile came from the blind side of the *Stark* where the radar was blocked by the superstructure (M. Moore, 1987). There were also reports at the time that the Phalanx was not even operational at all, although the Pentagon denies this. The *Stark*'s captain blamed these technical failures for the death of his 37 crewmen (M. Moore, 1987; Associated Press Service, 1987).

In 1991, two suits by the relatives of dead crewmen against General Dynamics, the manufacturer of the Phalanx, were dismissed on grounds of national security. Before this cover-up a number of documents were released to the press. They indicate that it was technical problems with the Phalanx that led to the tragedy. Testimony showed that the Phalanx was on and that the missile that hit the ship did not come in on its blind side. General Dynamics engineers and *Stark* crewmen revealed that below-standard circuit cards were used, which meant that the Phalanx system could only run 6–12 hours before breaking down. In 1986, Navy tests indicated only 71 percent reliability with the Phalanx. Several General Dynamics employees admitted that they had used "outlaw" software to override the Navy's circuit-testing program, and they even turned over one such program to the Navy. David Villanueva, a senior engineer for General Dynamics, said in a sworn statement that "Workers, including myself, often discussed our hopes that the Phalanx would never be relied upon by the Navy during a critical military engagement." Thomas Amile, a top Air Force radar expert, said in a sworn statement that, "the reliability of the Phalanx systems deployed on board U.S. Navy ships is extraordinarily low." The former *Stark* captain, Glenn R. Brindel, concluded that his ship "could have . . . avoided the tragedy . . . if the shipboard systems had performed to their represented capacities" (Webb, 1991).

The Navy, however, blamed the captain. This was certainly on the mind of every U.S. Navy captain in the Persian Gulf from then on, especially Capt. William Rogers III of the Aegis cruiser USS *Vincennes*.

The Aegis system is, as Congressman Charles E. Bennett (Dem., Fla.) once said, the U.S. Navy's most advanced "shipboard defensive system against attacks from aircraft and sea skimming missiles" (Congress of the United States, 1989, p. 15). It is probably the most advanced such system in the world, and it is almost certainly the most complex and advanced computer system that has seen significant military action. In some respects it is an expert system; in others, an autonomous weapon—in either case an information and weapons control network of great sophistication.

The Aegis system takes in various kinds of electronic information (radio and radar signals from other sources; radar reflections from the ship's own radars; the ship's own position; target and other information from humans) and distills it into information for the humans and the weapons through a system of 16 Unisys UYK-7 mainframe computers, 12 Unisys UYK-20 minicomputers, and scores of terminals, run by 28 different computer programs. The weapons include Harpoon ship-to-ship missiles, Standard ship-to-air missiles (up to 122), and two 6-ton, 6-barreled Phalanx automatic machine guns that can spit out fifty 20-millimeter uranium-core bullets a second. The humans can either choose the targets and tell the computer system to engage them or put Aegis on automatic and it will do all that itself (Editors of Time–Life, 1988, pp. 94–96).

There are actually four Aegis operating modes: automatic special, automatic, semiautomatic, and casualty. Only in the automatic special mode are targets meeting specific preprogrammed criteria automatically fired upon. Even in this mode humans can manually override the system. All the other modes have humans in the loop to some extent. However, in these modes the Aegis system automatically inserts targets into the engagement queue and schedules equipment for launching and terminal illumination. Trial intercepts are computed in all modes "and a time of fire predicted" (Pretty, 1987, p. 160).

In an interview, Rear Adm. Wayne Meyer said, "My guess is that in a *Vincennes'* situation no admiral would ever have given any other ship the right to open fire" (1988, p. 103). In understanding the destruction of an Iranian Airbus Flight 655, on July 3, 1988, the cause most emphasized by the Navy was the context in which the *Vincennes* was operating. The very reason the Aegis cruiser had been sent to the Persian Gulf was intelligence that the Iranians were considering using Chinese Silkworm missiles to attack U.S. ships in the Gulf. Other information indicated that Iran was planning to strike a major blow against the United States (the "Great Satan") around July 4. There was also official speculation that the Iranians might arm F-14 fighter-bombers with air-to-sea missiles, instead of the iron bombs that they normally carry, or use a commercial airliner for a kamikaze attack. Despite these concerns the Navy did not, or could not, supply air cover to the Persian Gulf units, so that direct identification of air threats by carrier aircraft could not take place (Goodman, 1989). There was also the acute awareness of the *Vincennes'* crew of the crippling Exocet missile attack on the *Stark* the previous year, a major encouragement to be aggressive.

Finally, and most important of all, the immediate situation of the *Vincennes* was crucial. She was several miles inside Iranian territorial waters engaged, with the frigate USS *Elmer Montgomery*, in a surface action against a number of small Iranian patrol craft, one of which may have been sunk by the *Montgomery*. Despite initial U.S. claims that this battle took place in international waters, later admissions by the U.S. government in the World Court, where Iran sued over the incident, and by Adm. William Crowe (who was serving as Chairman of the Joint Chiefs in 1988) on national television (July 1, 1992) make it clear that this was not so. There is even strong evidence that the United States was involved in a conscious attempt to provoke the Iranians into an attack on the U.S. units (Barry and Charles, 1992).

This surface action was just starting when the seven minutes of Flight 655 were tracked by the *Vincennes*. Also, one of the automatic 5-inch guns on the ship was jammed and so, in attempts to free it, the ship was "maneuvering radically" (Friedman, 1989, p. 74). There was also an Iranian P-3 aircraft flying nearby in a pattern it might well use if it were directing an attack on the *Vincennes*.

Relying mainly on official and semiofficial military accounts (Congress of the United States, 1989; Friedman, 1989) it is possible to set out in a rough chronological order the mistaken assumptions and judgments that led to the downing of Flight 655.[12]

1. One minute into the flight of 655 the identification (ID) supervisor on the *Vincennes* decides that the target blip wasn't Flight 655 because it was 27 minutes past the scheduled departure time. On the Aegis display, Flight 655 is indicated at this time by a symbol that means unidentified, assumed hostile aircraft.

2. The ID supervisor sees a military Mode II IFF (identification friend or foe) signal. Aegis data shows no record of this "squawk." The best theory of its origin is that as Flight 655 moved away from Bandar Abbas airport the IFF operator failed to move the gate that searched for IFF signals with it right away. So it accidentally picked up a stray Mode II signal, probably from Bandar Abbas airport, which had military as well as civilian functions and where F-14s were stationed, instead of the Mode III civilian–military signal that Flight 655 was broadcasting and which was picked up by many others, including the USS *Sides*, a frigate operating nearby.

3. Aegis picks up Flight 655's civilian IFF signal of 6760, the human operator reads it as 6675. Someone says its an F-14. The civilian IFF signals aren't known to the crew of the *Vincennes* anyway.

4. Responding to an unknown voice on the internal communication system an operator tags Flight 655 as an F-14 on the tactical display. From now on Flight 655 has a small notation "F-14" next to its symbol.

5. The combat information officer says that the target is possibly a commercial plane, but at the same time the antiaircraft officer, code named Golf Whiskey, is reporting that the target is descending and accelerating (faster than a commercial airliner could go) toward the ship. Even though the Aegis record and the observations of all other ships show that Flight 655 was ascending and at a moderate speed, at least two other *Vincennes*' crewmen also claim they saw information that Flight 655 was descending and accelerating at more than one instance (Congress of the United States, 1989, p. 30).

6. Even though the USS *Sides* had correctly identified Flight 655 as a nonthreatening commercial airliner, her captain did not argue with the *Vincennes'* labeling of the target as an F-14. Here is a quote from Rear Adm. William Fogarty on why this was so: "His ship was not being threatened, which it was not, and that the *Vincennes*, being a very capable AAW ship, if they call an F-14, then they are probably right" (p. 21). During this time there was "growing excitement and yelling" of "com air" in the combat information center of the *Sides*, which indicates to me that at least some of her crew realized that the *Vincennes* had made a mistaken identification and was about to shoot down a civilian airliner (p. 45).

7. Because of the data analysis Aegis performs automatically there was no way for the operators on the *Vincennes* to evaluate the radar blip directly, which might have allowed an experienced radar operator to deduce that it was not an F-14 (Friedman, 1989, p. 78; Congress of the United States, 1989).

8. Because of their need for enough distance to allow for full acceleration, the Standard missiles had to be used or the *Vincennes* would have had to rely on the one Phalanx automatic machine gun that was working. The other one was apparently out of commission (Friedman, 1989, p. 74). The Phalanx is a weapon that many in the U.S. Navy apparently don't trust on automatic. It also failed to spot the missiles that hit the *Stark*. That two of the *Vincennes'* close-in weapons, both automatic systems, were not working during combat is something that has not been remarked on in many of the reports. Add in the *Stark's* experience, and it is not a good showing for automatic gun subsystems.

Clearly, many scientists, military officers, politicians, and bureaucrats believed the Aegis would be effective in combat. However, some critics had warned that such highly automated defense systems would be far more trouble-prone than their proponents anticipated. One important distinction must be stressed—*Aegis is a man–machine weapon system*. That the human parts of this system committed most of the crucial errors is interesting but not an argument that the system didn't fail.

All of the failures of the Aegis system so far in combat were predicted by computer scientists. They also represent a case study of the various dangers of deploying so-called intelligent computer systems in military contexts. Moreover, there was the total inability of training and/or modeling simulations to predict the combination of decisions that led to the downing of Flight 655. That the crew had gamed threatening scenarios again and again for nine months, but not the harmless passage of a civilian airliner, undoubtedly contributed to their misperceptions.

As one analyst put it:

The reality of the nine months of simulated battles displaced, overrode, absorbed the reality of the Airbus. The Airbus disappeared before the missile struck: it faded from an airliner full of civilians to an electronic representation on a radar screen to a simulated target. The simulation overpowered a reality which did not conform to it. (Der Derian, 1990, p. 302)

The most significant single cause of the shooting down of Flight 655 seems to be that in the stress of their combat situation the crew of the *Vincennes* interpreted all information as confirmation that Flight 655 was a threat, even when that involved directly misreading the data supplied by the Aegis. Aegis itself provided the rationale for sending the *Vincennes* into the Persian Gulf, and because of Aegis she was given very liberal rules of engagement. Since *Vincennes* had the Aegis, the *Sides* did not challenge her identification of Flight 655 as an F-14. Having the Aegis gave the captain and crew of the *Vincennes* the confidence and ability to shoot down the Iranian airliner, but not the ability to see it was actually nonthreatening, nor the confidence to risk that it was hostile.

That computers can change human decision making has been ignored by many analysts, although there is a growing understanding of its importance (Winograd and Flores, 1986). In military terms it has already led to a number of significant errors:

The preoccupation with computerization dominates military technology, which in turn dominates military thinking, which itself dominates foreign policy, i.e., international politics. This is because our political leaders, our government bureaucracies, and much of the general public take for granted that superior computer technology guarantees superior military technology. In turn it is assumed that superior military technology guarantees superior military power, and that superior military power guarantees political effectiveness on the international scene in the pursuit and protection of our national interests. (Ladd, 1987, pp. 298–299)

The Aegis gave the *Vincennes'* captain and crew the illusion that they knew more than they did. It is an illusion that is not unique. Writ large it is U.S. military—and therefore technoscience—policy.

Chapter Four

The Uses of Science

By accepting the machine as his model, and a single
unifying mind as the source of absolute order, Descartes in
effect brought every manifestation of life, ultimately, under
rational, centrally directed control—rational, that is,
provided one did not look too closely at the nature and
intentions of the controller. In doing so, he set a fashion in
thought that was to prevail with increasing success for the
next three centuries.
—*Lewis Mumford (1970, Vol. 2, p. 98)*

Consciousness is an overrated concept.
—*Marvin Minsky (1989, p. 58)*

Without science, contemporary war would be impossible. And as
machine metaphors dominate war so they do science, although now the
machine is clearly a computer, an information processor. This is probably the
central theme technoscience has reinforced in the military, and in its
variations this is the main point of this chapter. But it is complicated. At the
heart of most dreams for absolute information there is the ideal of pure
intelligence, of artificial intelligence. It is a pervasive and peculiar version of
ration-ality that is masculine, mathematical, emotionless, and instrumentalist.

To understand late-twentieth-century war is to understand how this
rationality has shaped technoscience. Starting with the very specific
subfield of artificial intelligence, this chapter will track through various
definitions and metarules of working science to try to reveal something
of how they shape war's discourse and why technoscience, war, and
certain specific (and currently dominant) versions of rationality share
such an affinity. Since the situation of postmodernity frames technoscience now, a discussion of science's role in various versions of postmodernism ends the chapter.

Definitions of Artificial Intelligence

The dream of AI, the formalization/mechanization/creation of intelligence, goes back a long way. Ancient creation stories are often about some divinity's artificial intelligence project and its problems. Even the human quest for artificial intelligence predates science by thousands of years, witness Pygmalion's Galatea (made for him by Aphrodite, goddess of love) and the cabalistic tales of the golem. Certainly the idea is older than G. W. Leibniz and his perfect mathematical language and René Descartes' perfect machine. It is at least as old as Plato with his perfect models, models he thought were more real than real, more real than us, shadows on the cave wall.

Ramon Llull (or Raymond Lully), Thomas Bacon, and many alchemists of gnostic, Moslem, Christian, and pagan bent have had their talking heads or sought the creation of the homunculus, the artificial man. AI has not always been linked to the demands of war, but the AI dream today is a militarized dream, and in a strong sense it is a dream of an impossible human future, where war is a science. There are other equal if more subtle dangers from militarized AI.

Paul Edwards points out that cybernetic work is one of the most important touchstones of human self-definition of our times. As such it represents a profoundly significant political and psychological focus.

> Cybernetic science has become a primary *source of meanings* for us. Formal/mechanical modeling practices are therefore political issues in the profound sense that they provide the categories and techniques by which we understand ourselves. (1986b, p. 39; emphasis in original)

Just exactly what AI means is fluid. One common definition is that if it takes intelligence to do something by a human, then if a machine can do it you have an artificial intelligence. However, this misses actions humans can't do—actions that if we could do them we would certainly consider intelligent, such as distinguishing thousands of ICBMs from decoys and then telling hundreds of weapon systems when and where to shoot them down, all within a few seconds. These are all feats that AI researchers are trying to get machines to do. Some work originally done in the field, such as time-sharing, is now considered mainstream computer science, leading to the joke among computer hackers that "if it works, it isn't AI."

Specific subfields of computer science that are conceded to fall within AI are expert systems, natural language processing, and most learning and reasoning research. AI also generally describes work involving symbolic processors—computers that use logical inferences of one kind or another, as opposed to the strictly binary mathematics used in traditional programs and machines. The trendy second wave of neurocomputing sometimes tries to distance itself from AI, but its focus on modeling human neural nets and its specific history place it firmly in the AI discourse system. Robotics is also a closely related subfield, as is all sophisticated human–machine interface work, especially virtual reality research.

For this discussion AI will be defined as any artifact (software, hardware, or both) that performs a task that if done by a human would be considered as requiring intelligence. This does not mean that humans can or do perform the task. Locating a target and firing missiles at it would be considered AI under this definition.

Within the discipline of AI there are many different perspectives, even divisions. One distinction often mentioned is between those who model their work on human intelligence and those whose main emphasis is results. While this is an important difference that sometimes corresponds with a line between applied AI research and more theoretical work, there is a much clearer border between the AI research approaches that define intelligence as limited to specific terms of logic and formal constructions and those that don't.

Most AI researchers subscribe to the idea that human intelligence is not very complicated and that it will soon be completely modeled in formal logical systems. They may differ on the metaphors they use, the terms of the "frames" (matrices of related data), "scripts" (the regular patterns of typical and limited domains of data), or some other label; but they are committed to the idea that everything important can be described in words, the words linked logically with numbers, and the results determined inevitably by the ideal "rational man" or computer.[1] They do not accept the insight from the *Bhagavad Gita* that "words turn back" from some truths, nor the idea put forward by feminists especially that intelligence is very complicated and situated involving emotions and embodiment instead of just linear reason (Haraway, 1988).

Instead, some very powerful AI scientists at such leading centers as Carnegie–Mellon, MIT, Stanford, and Berkeley explicitly are working for exactly unsituated, disembodied intelligence through research on pure AI and on "downloading," putting a specific human's consciousness into an artificial brain (Fjermedal, 1990). These AI scientists are among those who have been making the numerous and unmet predictions for great break-throughs in AI for the last 30 years (Dreyfus, 1979), and they continue to do so (Nilsson, 1995). This dominant AI view is what Paul Edwards (after John Seely Brown) has called the "AI mentality" and which could just as well be called the AI dream or the AI fantasy. Edwards describes it as

> an approach to inherently vague, ill-defined, constantly changing prob-lems. Its central belief is that by circumscribing a specific problem domain and formalizing its features, problem-solving algorithms and heuristic logics can be created to find solutions automatically. . . . As a significant scientific, political, and social discourse, AI did not emerge until the Second World War, when technologies were invented that made it a practical possibility. (1986a, p. 1)

In the military it has been named Digital Realism. Fred Reed likens it to a species of Magic Bulletism, what he terms a belief in a perfect weapon. He goes on to explain that

> There is a powerful, almost universal desire to envision wars as predictable, precise, and bloodless. Digital Realism is the chief intellectual tool for doing this.
> It consists in believing that the world is like the inside of a computer. In programming a computer, all things are clean and certain. Each instruction does one thing, precisely described in the manual, with only one possible result, which can easily be ascertained. As long as a program is quite small, it will run to a foreseeable end with no surprises, click, click, click. This is a world of godlike certainty. (1987, p. 40)

Within the AI research community, there are strongly articulated alternative viewpoints to this paradigm. They emphasize how complex thinking is, and they argue that human language use is based on the existence of a discourse community and on the interactional basis of communication. To reveal how different this view is from the dominant AI mentality will take some explaining. The first step is to look at science itself.

Definitions of Science

> The man of science is not content with what is found on the surface of the earth. . . . [He] has penetrated into her bosom . . . for the purpose of allaying the restlessness of his desires or of extending and increasing his power. . . . Science has bestowed upon the natural philosopher powers which may be called creative; which have enabled him . . . by his experiments to interrogate nature with power, not simply as a scholar, passive and seeking only to understand her operations, but rather as a master, active with his own instruments.
> —Sir Humprey Davy (quoted in Easlea, 1980, p. 248)

Any definition of science is now contested. There is more debate today about the science question then there has been since the sixteenth century. There isn't space here to give a full account of the various perspectives; instead, this section will try to point out aspects of this struggle over science's meanings that seem important to understanding postmodern war.

The most important of these disagreements is the debate between, to but it crudely, constructivists (sometimes called relativists) on the one side and realists (who overlap a great deal with positivists) on the other. There are those among the latter group who claim there is a direct and clear connection between science and reality. Some of them may admit that society and human perception can muddy the relationship at times, but in their view science is always the best way of understanding the world. Maybe, they argue, it is not a perfect system, but it is far superior to any other. While definitions of science differ wildly among this group, they all specify certain values. These metarules of science are logical rationality, experiential objectivity, and physical materiality. Sometimes there is also an emphasis on repeatable experimentation or regular intervention on matter. A few philosophers, such as the purest of the logical positivists, depend on elaborate definitions of the logically provable or falsifiable to define science. But, by and large, it is their own reason, their own fairness, and their own existence that most scientists assume. This is the dominant point of view.

This privileged view of science has been challenged by the constructivists in a number of different ways. Some have undertaken painstaking personal studies of scientists and the construction of scientific facts in order to show just how culturally determined and somewhat arbitrary this process is. These studies often rely on detailed textual and discourse analysis, as well as historical and other interpretive approaches besides the anthropological assumptions that underlie much of this work.

Some of these researchers even go so far as to claim that science, as practice and social object, is entirely constructed, and that there is no direct, or even discussable, connection between science and the "real" world, a world which might not even exist, after all. Most constructivists do not hold such absolute views, but some do.

However, most who argue that much of science is socially defined would never claim that all of science was socially determined. Helen Longino has one of the best summaries of this bounded constructivism:

> There is a world independent of our senses with which those senses interact to produce our sensations and the regularities of our experience. There is "something out there" that imposes limits on what we can say about it. Once we have decided on a system for measuring movement, the speed of an object is not arbitrary. The sorts of things we measure, however, will change as our needs, interests, and understanding change. The processes that occur in the world will continue regardless of changes in our descriptive systems. . . . The fact that not all changes and gradations can be encompassed by any given system and that the changes and gradations that are important to us change over time is one of the phenomena driving scientific change. (1990, p. 222)

Several theoreticians have explored this zone between realists and relativists. Sal Restivo says he would rather be considered "realistic." He advocates a "wedding" between a constructivist view of science and a realist view of the world (1988). In this view, science has access to the world through some of its methods, but which parts of the world it explores and reshapes is determined socially and personally, not "naturally."

Donna Haraway says something very similar when she criticizes both relativism and totalizing versions of realism and logical positivism:

> The alternative to relativism is not totalization and single vision, which is always finally the unmarked category whose power depends on systematic narrowing and obscuring. The alternative to relativism is partial, locatable, critical knowledges sustaining the possibility of webs of connections called solidarity in politics and shared conversations in epistemology. (1988, p. 584)

Longino herself has charted in detail the ways science is configured as social knowledge. She demonstrates in several ways how contextual values can shape practices, delimit questions, affect descriptions, mold assumptions, and motivate acceptances in ways that determine what science comes into being. In this view science, and all reasoning are crucial, but limited, social practices. She remarks that "Treating reasoning as a practice reminds us that it is not a disembodied computation but takes place in a particular context and is evaluated with respect to particular goals." Finally, she stresses that experience is "an interactive rather than a passive process." Experience is a complicated product of "the interaction of our senses, our conceptual apparatus, and 'the world out there' " (1990, pp. 214, 221).

An insight that is usually closely related to a refusal of simple realism or simple relativism is the realization that objects, natural or artificial, are not passive objects-of-knowledge; rather, their specificity shapes possibilities. Nor can the scientist be just an observer; the specific approach of the scientist shapes possibilities as well. Stephen Toulmin describes this realization as postmodern:

> As we now realize, the interaction between scientists and their objects of study is always a *two-way* affair. . . . In quantum mechanics as much as in psychiatry, in ecology as much as in anthropology, the scientific observer is now—willy-nilly—also a *participant*. The scientists of the mid-twentieth century, then, have entered the period of postmodern science. (1982, p. 97; emphasis in original)

Bruno Latour (1991) has taken this idea beyond usefulness when he argues that "nonhumans" such as machines and microbes should have rights as humans do.[2] He does well to stress the importance of such nonhumans in science and how crucial laboratories and relationships of power and discourse are to the construction of science, but his conflation of human responsibility and inhuman objects not only doesn't explain how social knowledge is constructed but implicitly supports the politics of science as it is.

A stronger and more useful way of explaining the importance of artifacts is to see them as playing special roles within a discourse. In science, artifacts and experiments are powerful metaphors, arguments, sometimes even rules, of the discourse. They are never, however, acting subjects. Within the workings of discourse, then, a machine or an experiment is not neutral—it has "politics," for example (Winner, 1986). It influences reality by influencing how people think and decide (Aronowitz, 1988). In sum, then, science can be seen as both a discourse and a practice mediating between human culture and the real world on physical, psychological, and even emotional levels.

Seen in another way, however, science is a religion, an ideology, even a cult. Although other social systems can manifest powers of demonstration (miracles), prediction (prophesy), and experience (the subjective one of the believers), none approach science in its demonstrative, predictive, and experimental powers on the physical plane—that part of reality most easily visited (captured?) by science. Science kills. Science saves. Science intervenes continuously in our lives.

This stress on intervention as characterizing science is developed in the philosopher Ian Hacking's work. He points out that the central characteristic of science is the ability to make regular "interventions," which can be experiments or observations that produce regular phenomena (1983). The theoretical side of science, representation, not only tries to explain these interventions, in some cases it can organize them as well. Thus the phenomena are produced by both the theoretical and actual human intervention into the material. The process continues as representations influence interventions and vice versa.

This is the methodology of AI: proof by production not by repeatable experiment. It is also, by necessity, the way a science of war must proceed, for every battle is different. But what must be produced as proof? Victory? Security? A strong military? War? The confusion is because experiments don't produce proofs so much as definitions. The science of war defines wars, but it can't promise to win them.

There is something potent in science. There is powerful magic in the regularity with which science can mine (find, define) facts. The power of nuclear weapons, moon trips, and open heart surgery has meant a very special place for science in modern culture. This effectiveness of science may well be integral to human nature and culture as they have evolved up to this day, having more to do with survival than with truth. Science works on more than it explains—which explains some of the ways it has shaped war.

Even if a scientific (experimental, pragmatic, reductionistic) mentality isn't a necessary part of being human today, science is just too central to be abolished without cataclysmic changes. The solution is not just to wish science away but to change it.

In this view the world, our world, is an experiment, not just a definition or a demonstration. Ian Hacking has persuasively argued that

> Experimental work provides the strongest evidence for scientific realism. This is not because we test hypotheses about entities. It is because entities that in principle cannot be "observed" are regularly manipulated to produce . . . new phenomena and to investigate other aspects of nature. (1983, p. 262)

Experiments such as this are really interventions. It is sometimes hard to remember that science is just a subset, albeit an important one, of knowledge, though almost everyone calls this wisdom when they say it. However, with most of the small details of our daily lives we choose science.

To understand this mutual shaping, the relationship of the techno-science of AI and doctrines and practices of postmodern war, it becomes necessary to wonder about science's real powers and its limits, and how they are entangled. This attempt to undo a few of these knots is based on the premise that science is both social construction and social constructor. Performing the Turing test or similar experiments changes the definitions of concepts central to our own self-image. We can expand our definitions, or limit them, or change them, but we can't keep them unaltered. Science is not normative, it is intersubjective. While science is undeniably powerful, it is just one such system. It works, but in strange ways, not just on nature but on ourselves. And on our idea of nature. And on our idea of ourselves.

Rationality in Many Guises

Certain models of rationality based on formula, calculation, and logic are at the center of science's claims. These models are the same ones that make up the core of the AI paradigm. The same versions of rationality are a key part of the epistemology of military science, and they are a central theme of the discourse of war. Many philosophers and others question these claims of rationality, even from within science.

Computer scientists Terry Winograd and Fernando Flores, for example, distinguish between the rationalistic tradition and rationalism itself. They argue that the rationalistic tradition, with its emphasis on formal logics and mathematics, "often leads to attitudes and activities that are not rational when viewed in a broader perspective" (1986, p. 8).

The AI paradigm is part of this rationalistic tradition writ into computer languages and human–computer discourse. Much of it is framed in assumptions of decision science and the "rational man" popularized by Herbert Simon. The connections between the rational man of economics, the artificial man of AI research, the pure scientist, and the perfect soldier are numerous. All these creatures share a common heritage. At their core are the same metarules of individuality, rationality, masculinity, and domination. All of them are supposed to rely on logic, not emotion, for solving their problems. Relying on the "intellectual technologies" of systems analysis, management science, and related techniques to measure the costs versus benefits or economic expected utility, decision scientists claim they have created "a social-scientific approach to decision making." Have some knowledge engineers put it all into a machine and you have an expert system.

In her analysis of Simon's work, Carolyn Miller (1990) reveals how it is focused on the metaphor of rationality in both economics and AI. Economic expected utility, complexified to multiattribute analysis and rational decision making, is now called problem solving. This has been Simon's focus for fifty years, although he conflates it to "The nature of human reason. . . . " Miller goes on to show how Simon's idea of rationality is "instrumentalist," in that Simon has replaced "Olympian Rationality" (omniscient and absolute) with procedural rationality. She calls this "scientistic" and contrasts it to her idea of reasonable rationality, that is, "the discovery and articulation of good reasons for belief and action." So, in Miller's terms, history, convention, insight, emotion, and value all become rational, that is, possible "good reasons" for thinking or doing something.

History? Insight? Emotion? Value? These elements are difficult to formalize, to put it mildly. Unsurprisingly, vulgar rationality remains dominant in many guises in many related discourses: the short-term rationality of our politico-military affairs with its roots in economic metaphors of material gain and loss (Chomsky, 1986) and the limited rationality of the bureaucracy, the computer, the man-on-the-street.

Which isn't you or me. He is the computer model of the average consumer, voter, soldier, investor, or enemy; he is the rational man of economics, the "rational opponent" Henry Kissinger calls him. Kissinger defines him as someone who "must respond to his self-interest in a manner which is predictable" (quoted in Falk, 1987, p. 16). Being predictable by Henry Kissinger is rational? Some feminist philosophers have argued it is only a limited type of rationality, and it is predicated on relationships of domination. To show some of these relationships clearly and with their deeper structure, it will help to look at the critique of the role of limited versions of rationality in science that comes specifically from Carolyn Merchant and Sandra Harding.

Merchant (1982) draws convincing lines between limited conceptions of rationality that are central to science and an attitude and practice toward nature of domination, dismemberment, even destruction. Given the connections expressed by the founding fathers of modern science, men such as Francis Bacon, it is not a hard case to make. Merchant shows how much Bacon relied on the metaphor of torturing witches to present his vision of science's relationship to nature. To quote Bacon himself in a passage addressed to his king, James I, and referring to the king's famous persecution of witches:

> For you have but to follow and as it were, hound nature in her wanderings, and you will be able when you like to lead and drive her afterward to the same place again. Neither am I of the opinion in this history of marvels that superstitious narratives of sorceries, witchcrafts, charms, dreams divinations, and the like, where there is an assurance and clear evidence of the fact, should be altogether excluded. . . . Howsoever the use and practice of such arts is to be condemned, yet from the speculation and consideration of them . . . a useful light may be gained, not only for a true judgment of the offenses of persons charged with such practices, but likewise for the further disclosing of the secrets of nature. Neither ought a man to make scruple of entering and penetrating into these holes and corners, when the inquisition of truth is his whole object—as your majesty has shown in your own example. (quoted in Merchant, 1982, p. 168)

Bacon was only echoing a common theme. Sir Davy Humprey, one of the first presidents of the Royal Society, has already been quoted at the start of this chapter. In his book *Witch Hunting, Magic and the New Philosophy*, Brian Easlea (1980) has dozens of similar quotations from the founders of modern science. These new philosophers often used images of torturing and forcing nature (the experiment), in contrast to the approach of the scholastics, followers of Aristotle, and the alchemists, who wanted merely to enter into a conversation with nature. To further distance themselves from the alchemists, whose rituals often resembled experiments, the new philosophers stressed the absurdity of the alchemists' belief in impossible occult forces, such as the moon pulling on the tides.

Robert Boyle noted that there were two reasons for studying nature: "For some Men care only to Know Nature others desire to Command Her" (quoted in Easlea, 1980, p. 247). Commanding is easier then knowing. Easlea points out that the experimentalist natural philosophers rejected the persecution of witches on the grounds that the occult didn't exist, there was no magic. They believed this because their metaphysics was material and based on Descartes' machine metaphor. Instead of there being witches, nature would have the role of witch, and she would be tortured for knowledge and power. "Knowledge is power" is Bacon's most famous insight. Or, as Foucault (1986, p. 236) put it, "Knowledge is not made for understanding; it is made for cutting."

The word *science* does probably come from "to cut," and *knowledge* traces its meaning from "to be able." *Judgment*, on the other hand, has its roots in "showing" and "demonstrating." Reason and rationality are directly from "calculating" and "to count." *Science* is from the Latin *scire*, "to know," probably originally "to cut through," maybe from Sanskrit *chyati*, "he cuts" and Farsi, *scian*, "a knife." *Knowledge* derives from a whole set of Anglo and Germanic words for "can," and "to be able." *Judgment* is from the Old Latin *deicere* and the Greek *deiknumai*, "to show" and the Sanskrit *disati*, "he points out or shows." *Reason* and *rationality* are directly from Latin's *ratus*, "to count or calculate" (Partridge, 1966). These all can be seen as different epistemologies. Today, the meanings overlap a great deal, but perhaps something could be gained by drawing some distinctions between them. Humans have many different criteria for thinking, for proof as well. Certain styles of thought have always been with us. Consider how the belief that numbers alone hold the key to understanding goes back, from the twentieth century (Porter, 1995) to the ancient Greek Philolaus, who proclaimed:

> For the nature of Number is the cause of recognition, able to give guidance and teaching to every man in what is puzzling and unknown. For none of existing things would be clear to anyone, either in themselves or in their relationship to one another, unless there existed Number and its essence. (quoted in Heims, 1980, p. 60)

As modern science developed out of the philosophical debates and social realities of the Enlightenment, it was shaped in many cases by historical chance as much as design. Distilling the idea of experimentation from alchemy, the new scientists went to great pains to reject alchemy's metaphysics, so strongly keyed to female-positive images (e.g., Sophia, goddess of wisdom; the respectful partnership with nature that was gendered female; the symbolism of hermaphrodites and androgyny). The new philosophy, now known as science, would be a masculine project in two senses. First, adapting logic and emotionless calculation from the scholastics and marrying them to objective clinical experimentation. Second, in its rhetorical equation that woman is nature and science is torture and domination, as Bacon says so clearly, which makes of the scientist an inquisitor.

This gendering of nature—and the metaphor of domination explicit in the way science uses it—is an important part of the power/knowledge of military AI. Central aspects of military discourse are also gendered in the key of domination, as is war itself. So the affinity between science and war that has become so obvious in the twentieth century should not be a surprise. They share similar metarules about gender, rationality, instrumentality, nature, and domination. They also have extreme differences. And it is important to stress that within both discourses there are other views besides the dominant ones. But still, the many similarities between war and science are striking.

Science has changed in many ways since Francis Bacon's day. Now scientists do experiment on and with occult (hidden) forces that were once considered by most scientists to be impossible. But science still remains a form of limited rationality constrained by the material limits of intervention and representation. If science claimed no more, that would be one thing, but instead it often wears the guise of sweet reason itself.

Some critics and scientists have called for integrating science with the rest of society, for scientists discarding the cloak of impartial rationality and accepting some social responsibility. Sandra Harding goes further and elabo-rates a postmodern feminist revaluing of science to make "moral, political, and historical self-consciousness . . . of primary importance in assessing the adequacy of research practices" (1986, p. 250).[3]

How is this intervention described and evaluated? It depends on the rules. Many of the rules seem to be physical laws; many clearly are not. Which rules are applied depends on the metarules, the rules about the rules. What counts as evidence? Who is allowed to speak? To publish? To make jokes? Who is made into a joke? To approach these questions the figure of postmod-ernism should be sketched more fully because, above all, postmodernism seems to mark a growing concern with these metarules or at least a growing difficulty in hiding them. And the game of discourse that is being ruled and metaruled must be discussed in more detail as well, with specific attention to the conversations of the AI community.

Postmodernism

Since the mid-1930s there has been talk of *post*modernism. Some postmod-ernists are defined only through their critique of modernism. Deconstruction and sometimes poststructuralism are often put in this category. By no means is there a consensus among philosophers about the meaning or value of postmodernism (let alone deconstruction or poststructuralism or anything, actually). However, beyond philosophy there is more clarity. In certain fields "modern" has taken on very specific historical and/or symbolic meanings. Architecture and art are the best examples. In other areas, such as the history of war, modern war is defined both chronologically and qualitatively.

Postmodernism also has varying qualities in different domains, but there do seem to be certain themes that appear with regularity as they do with science:

- Information is the organizing principle. Science, war, politics, and business all situate information at the center of their particular discourses. At the same time there is a disillusionment with the grand master narratives of modernism: art, progress, science, technology, democracy, and nationalism—although technoscience has grown in cultural power until it dwarfs the others in relative terms. But there has been a breakdown of the modernist view that science is value free; in postmodernism the demonology of science coexists with its ritual adoration.

- Knowledge is seen as situated, specific, not always generalizable by some; at the same time numerous others believe in potentially hegemonic epistemologies, almost all of which are at odds with each other. Yet, there is order. The discourse of "natural" rules, mediated by science and filtered through rationality, oversees a domination of no domination. The market, the battle, and the scientific method determine most of the choices offered in the dominant discourse.
- There is an emphasis on the metarules over the rules of particular systems.
- There is a tremendous increase in speed, and the Cartesian grid is now a Cartesian box.
- Human–machine culture proliferates.

These themes all appear with a vengeance in postmodern war. To understand how this is a break with ritual, ancient, and modern war will require a long detour into history.

Part Two

The Past

An experiment in artificial holocausts. The incendiary bombing of Hamburg, Germany, was a test of the practicality of creating artificial firestorms. The success of the Hamburg raid, and especially the destruction of Dresden in a later experiment, inspired the United States to make such "strategic bombing" a primary focus of the air attacks on Japan. This photo is from June 20, 1944, long after most of residential Hamburg had been burned, and shows the results of a successful attack on the Ebano oil refinery by B-17s and B-24s of the 8th Air Force. USAAF files, NASM Library, photo 21435.

Conventional apocalypse: The firebombing of Tokyo. The artificial firestorm created during the incendiary raids on Tokyo killed more humans than any other act of war in history, including the later atomic bombing of Hiroshima and Nagasaki. This is an official U.S. damage assessment Tokyo bomb map. It clearly depicts the postmortem city and the literally inscribed body politic. Seen from the present, we note how our contemporary political economies insist that the old must make way for the new; destruction = construction. It is also clear that atomic weapons are merely the symptom of postmodern war, not its cause. Finally, even this death camp of a city is cyborgian, the clean rationalized (destroyed) parts coexisting with the organic, still living, sections. Tokyo City Damage Assessment Report No. 20. CIU XXI Bomber Command. National Archives.

Bomber glorioso. War is beautiful. An unidentified U.S. bomber flies above the storm through a cascade of sunbeams in World War II. High-technology weapons, and the threat of apocalypse they bring, has done nothing to eliminate the visual and other sensory seductions war offers. If anything, the awesome power of twentieth-century war machines to destroy in massive cathartic explosions and to transport us to extraordinary new locales of incredible beauty have made war even more attractive for some. USAAF files, NASM Library, photo 3328.

The brain of a human–machine weapon system. As the official description puts it: "China: Operations. Men plotting the course of 14th Air Force planes in the fighter control section of the 312th Fighter Wing, 14th Air Force at a base somewhere in China. Note aircraft status board in background. 14 August 1944." War is no longer a matter of courage, of morals, of ideology, or of luck, the military planners hope. It is a matter of technology wedded to humans in complex systems. The control of information is crucial. USAAF files, NASM Library, photo 2188.

The fist of a human–machine weapon system. Man–machine weapon systems were perfected in the twentieth century in many places such as this B-29 bomber cockpit. Specialized, trained, linked intimately with their weapons and weapon platform, these men of the U.S. Army Air Force had their very missions, and their lifespan, calculated out mathematically by men such as Col. Robert Strange McNamara, the ex-Harvard economics professor turned military analyst. While effective within limited domains, such calculations are irrelevant in the wider context of military strategy (as the failure of strategic bombing attests) and they are obscene in the case of the individual humans, called "inventory" in one of McNamara's reports, to which they are applied. USAAF files, NASM Library, photo 3252.

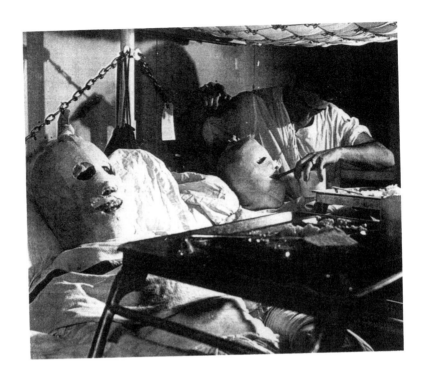

The human face of war. Modern and postmodern war have much horror to offer us, as these sailors, burned when a kamikaze hit their ship, have discovered. Despite its patriarchal ritualistic resonances, which still account for much of its appeal, war has become something that happens to us, not something that we do. Human–machine systems, bureaucratic and weapons, can influence and decide battles, but individuals are now only victims. Perhaps there is a stoic heroism in taking one's part as a living cog in the (post)modern war machine, but in the long run it is difficult to see how such heroics serve the future of humanity. Victor Jorgensen, USN photographer.

Searching the heavens. These questing searchlights are stabbing out of Tokyo looking for U.S. bombers to fix on so they can be destroyed. In World War II the air became a crucial, and much transected battlespace. Heaven and earth are here transfixed into killing zones. Battlespace has continued to expand and it now reaches into space, at least into low orbit, perhaps to the moon. On Earth as well, war has spread out to the cold of the arctic region and the deep sea depths and to every corner of the land, whether it is inhabitable or not. War has expanded in time as well, of course. These lights cut the night so aerial killers can be hunted. Vast industries were created to make this deadly hide-and-seek game possible; industries that dominate the postmodern economies of today. As war continues to multiply its modes and sites we must wonder how long before it is everything. USAAF files, NASM Library, photo 3121.

War meets nature. This is the culmination of World War II and the context, therefore, of postmodern war. As the official caption states: "Japan: Bombing, Nagasaki. A crumbled mass of roof tile is all that remains of a Japanese home near ground zero. Ground zero is the spot directly below the explosion of the atomic bomb. 14 Oct. 1945." Ground. Zero. Is this the calculus of the human future? Why not? USAAF files, NASM Library, photo 3686.

Chapter Five

The Art of War

The Discourse of War

> Discourse is discourse, but the operations, strategies, and schemes played out there are real.
> —Jean Baudrillard, in Baudrillard and Lotringer (1987, p. 15)

What can we possibly know about the history of war? Humans have fought each other in organized groups for thousands of years. We know bits and pieces of this history but we can't completely understand earlier wars. Why? Is it because they were part of cultures not our own? In part, but to my mind even the people there, at the time, couldn't fully understand the wars they fought. Culture is too complex for full understanding, at least by the humans who construct it as they live it. But even if full understanding is denied us, we can understand a great deal. Even more, we can seek to better understand contemporary war by looking at its antecedents.

This chapter is a quick and dirty tour of certain major themes in warfare that have become crucial today. It is hardly a complete history, and it is certainly full of contradictions because it focuses on the long-standing tension in war's discourse between the desire to understand, explain, predict, and control combat and the reasons why this desire is impossible to fulfill. Perhaps the best place to start is with the earliest known guide to war.

It was written in China over 2,300 years ago. The textual evidence indicates that one author with real military experience wrote the bulk of the 13 chapters now called—along with the extensive commentary added over the years—*The Art of War*. I'll refer to this author as Sun Tzu, since it doesn't make much difference here whether or not the book was written in the sixth century B.C.E. by a general for Ho-Lü, the king of the state of Wu, or by some other general sometime during the Warring States period (453–221 B.C.E.). In either case, *The Art of War* discusses the most important aspects of both ancient and modern war, the types of war which prevailed for more than 5,000 years up until 1945 (Griffith, 1962; Sun Tzu, c. 400 B.C.E.).

At the very beginning of his text, Sun Tzu argues that there are five "fundamental factors" in war: the moral, the weather, the terrain, command, and doctrine. Some of these factors are under direct human control, others are not. By Sun Tzu's schema those factors under at least partial human control include the logical (doctrine), the emotional (moral), and areas that are a mix of both (command). Throughout ancient and modern warfare the importance of these various elements remained in rough balance in the discourse of war, although they were called by many different names.

This balance was one reason war was traditionally labeled an art, not a science. While rationality in various forms was always considered crucial to successful warmaking, so were other forces, variously called the "moral" (Sun Tzu), "fortuna" (Niccolò Machiavelli), the "heroic" (Morris Janowitz), "friction" (Carl von Clausewitz), the "spirit" (T. E. Lawrence), or "intuition" (Col. Francis Kane).

Each of these military thinkers draws a sharp distinction between two distinct poles: the area of the natural (often described in terms of luck or human will) and the area of the rational (the logical, the planned). In his chapter called "Friction in War" Clausewitz explains:

> Everything is very simple in War, but the simplest thing is difficult. These difficulties accumulate and produce a friction. . . . Will overcomes this friction; it crushes the obstacles. . . .
> Friction is the only conception which . . . distinguishes real War from War on paper. The military machine . . . and all belonging to it . . . appears on [paper] easy to manage. But . . . it is composed entirely of individuals, each of which keeps up its own friction in all directions. (1962, pp. 77–78)

For Clausewitz it was the human that brought friction to the battlefield, as well as moral sense, intuition, and heroism. Although many herald Clausewitz as the ultimate theorist of contemporary war, his theories only explained war as it was, not as it was becoming. World War I and World War II took Clausewitz's theories, and modern war itself, past the point of absurdity. It is remarkable how Clausewitz is usually portrayed as a balanced, rational philosopher of war, while actually his emotional need for war was extreme even by his own account. Consider this letter (Rapaport, 1962, p. 22) to his fiancée, Countess von Brühl:

> My fatherland needs the war and—frankly speaking—only war can bring me to the happy goal. In whichever way I might like to relate my life to the rest of the world, my way takes me always across a great battlefield; unless I enter upon it, no permanent happiness can be mine.

After the war did start, he wrote another letter (p. 416) to her, right before the battle of Jena, which was a major defeat for the Prussians: "The day after tomorrow ... there will be a great battle, for which the entire Army is longing. I myself look forward to this day with joy as I would to my own wedding day." This is not a rational attitude toward war. Clausewitz's desire for war coexisted with his attempts to rationally explain it.

By the late-modern era this tension between a rational analysis of war and war's emotional appeal had become a dominant aspect of military culture. "The history of the modern military establishment," Morris Janowitz claims,

> can be described as a struggle between heroic leaders, who embody traditionalism and glory, and military "managers," who are concerned with the scientific and rational conduct of war. (1971, p. 21)

The emphasis in the military has historically been on rules and modeling because the reality of battle is biased in the other direction. For example, the military historian John Keegan points out that the fundamental purpose of training "is to reduce the conduct of war to a set of rules and a system of procedures—and thereby to make orderly and rational what is essentially chaotic and instinctive" (1976, pp. 18–19). But reason has never been able to make of war a science. It is impossible to completely quantify anything in war besides logistics and ballistics, let alone the emotionality of combat and the political shaping of war or peacemaking. Strategic models of rationality constantly aim for the rigid use of numbers and formulas (Walt, 1987). But reducing rationality to a system of rules and procedures may not be effective, or even rational, under many people's definitions of the term. Still, it has long been *a* goal of many military men and lately it has become *the* goal.

To understand the powerful attraction of formal systems that deny the emotional (including the moral and intuitive) and natural in war, it will be necessary to look more closely at war as a discourse system.

A discourse system is the communicative practice in a specific domain of knowledge, and its practitioners are sometimes called a community. The importance of discourse as a framework for understanding culture has been receiving growing recognition recently. A new field of discourse analysis has been marked out that includes work from the disciplines of linguistics, psychology, social psychology, sociology, anthropology, history, law, artificial intelligence, philosophy, mass communication, political science rhetoric, and poetics. What these very different approaches agree on is the central importance of language, as actually used in conversations and texts, for shaping not only a significant part of human understanding of the world but many material aspects of the world as well.

A discourse does not just involve words. For example, what warriors and soldiers do with their bodies is more important than what they say

(Scarry, 1985). Weapons, rituals, traditions, and techniques are all parts of the discourse of war. Over time the rules of war's discourse have shifted considerably, as have the hidden metarules that determine those rules. For example, such metarules, establish who can be listened to and who can't, as well as conventions that mark out the permitted areas for discussion and those forbidden, and rationales that allow certain questions to be asked but not others. Paying attention to the rules and metarules of a discourse system reveals a great deal about the people in the discourse community and the society as a whole that nurtures it (Foucault, 1980; Shapiro, 1981).

The analytic techniques used for looking closely at discourse are those available to any close reader; they are drawn from common sense and the long tradition of rhetoric. A close look at figures of speech, style, the rules of genres, (poetics in other words) offers a way to explore beneath the surface logic of military and scientific thinking. The technical details of the rhetoric of persuasion that are employed are clues about the assumptions, symbols, and myth systems that the creators of the discourse consider irrefutable and self-evident, even if only unconsciously. In etymology, in metaphor, metonymy, synecdoche, or irony, in genre, and in style there is evidence for some of what swirls beneath the surfaces of the discourse, no matter how official, how technical. The noted literary critic Northrop Frye explains: "The only road from grammar to logic, then, runs through the intermediate territory of rhetoric" (1957, p. 331).

A discourse has many meanings. In this version of war's history the focus is on the organization of the discourse system, its explicit, implicit, and latent rules, contents, and effects. To uncover the latent (the "dynamically unconscious repressed" as Hayden White [1973] has called it), a certain "psychologic" must be employed. Tropes (figures of speech) are a very important part of this analysis, representing as they do a swerve from the literal that often marks a point of emotional importance. Where there are powerful or numerous tropes and other stylistic devices it can be assumed that the story of the text (its argument or narrative) is being carried—at least in part—by implicit messages (psychological, symbolic) and not only by explicit logic.

In the case of scientific and government texts, uses of rhetoric are all the more significant because the style of the genre is itself the style of nonstyle. The language of the scientist, the engineer, and the official is supposed to be transparent. It is not supposed to become, or even influence, the content of what is asserted (Bazerman, 1981; Hofstadter, 1955; Medawar, 1951). Lately, this claim has been criticized from many points of view. Many scientists now recognize the importance of rhetoric, especially the use of metaphors, in the construction of scientific truths (Gould, 1987; Karush, 1986; Stent, 1986).

Even the most official document is communicating on many different levels: the official explicit content; the implicit policy and political implications; and more latent, emotional appeals, threats, and other expressions of the political unconscious, psycho-logic, the mass mind, public subconscious, or whatever one wishes to call it.

This way of thinking about war is not scientific. It seems quite likely that the complex reality this book addresses cannot be understood in a purely scientific framework. My analysis is meant to be suggestive, reasonable, even persuasive. This is a historical, philosophical, and literary understanding of a technical and official type of discourse about a bloody and awe-full reality. As such, you dear reader are free to take it as you will, even as an allegory.

The idea that war can be profitably understood as a discourse is a useful one. Primitive or ritual war has been brilliantly analyzed in terms of discourse by historian Susan Mansfield:

> Traditional cultures clearly understood war as a form of discourse between the human and "the other." As a result they accepted that the conversation must conform to certain syntactical rules and limits if the communication was to be effective. Moreover, they also hypothesized that war, like any other valued dialogue, was ongoing, continuous, and cyclical. The natural environment and the human enemy must, in some degree, be protected and preserved. (1982, p. 236)

Notice how important the enemy is, if war is to survive. Without an enemy there is no conversation. This is a realization that warriors have never lost, although in more modern times it has had to be disguised in the rationalization that the function of war is making peace. Actually, war has always been a conversation. This is even true of modern war, as the great scholar of war Quincy Wright was at pains to emphasize.

> Modern war tends to be about words more than about things, about potentialities, hopes, and aspirations more than about facts, grievances, and conditions. . . . War, therefore, rests in modern civilization, upon an elaborate ideological construction maintained through education in a system of language, law, symbols, and values. The explanation and interpretation of these systems are often as remote from the actual sequence of events as are primitive explanations of war in terms of the requirements of magic, ritual, or revenge. War in the modern period does not grow out of a situation but out of a highly artificial interpretation of a situation. (1964, p. 356)

Still, to realize war can be described as a discourse system doesn't explain all that much. Before going on with war's historical story it is necessary to explain how the discourse of war is also what Michel Foucault labeled a system of power and knowledge.

Power, Knowledge, and War

It is because war is so obviously concerned with power that Foucault's work is particularly relevant (1980). He argued that power is not just repressive; it also involves the positive shaping and control of discourse and the establishment of what he termed a "regime of truth." Much more important than knowing the details of the imposition of power is understanding how that power's very ground is organized and the unchallenged assumptions that make this organization possible.

According to Foucault there are several ways to examine a power/knowledge system: microhistories, analysis of its discourse, and investigation of its genealogy. Foucault takes the idea of a genealogy from Friedrich Nietzsche. A genealogy makes no claim to historical (even critical historical) completeness; rather it is a look at the "ancestors" of the current regime of power/knowledge with an aim to understand something of the forces that have shaped it. A genealogy works from the point of view of doing something today (*for* or *as* or *with* or *to* the descendant) based on what Foucault called subjugated knowledges. There is no real attempt to historicize the ancestors and see their world from their point of view. A genealogy is also, rhetorically, a rejection of any claim of being a purely scientific examination, which Foucault's earlier term "archaeology" implied.

Genealogy also signifies all the complexity of a living family, involving not just facts about occupations, travels, children, deaths, and so on but also the intangible elements of myth, mystery, incompleteness and the tangle of relationships wound with the emotions of love, hate, and indifference.

Foucault defines a genealogy specifically as the chronicle of a set of different knowledges, all marginalized in some ways from the dominant defining discourse. He breaks these subjugated knowledges into two categories:

1. Erudite knowledge hidden in historical details and exceptions that have been repressed by the dominant discourse and hidden by the broad generalizations of its myths and metarules.
2. Knowledges excluded because they failed some test of the discourse rules for a certain level of complexity, practicality, or formalization of logic or science

In the case of the current war system of power/knowledge the subjugated knowledges would include perspectives coming from marginalized and specialized forms of knowledge (especially from AI's subfield of philosophy, and within it the more critical perspectives such as phenomenology and hermeneutics); from practical experience low on the hierarchy (the worker's, the programmer's, the soldier's); from those who would raise political and moral issues; and from viewpoints rigorously excluded as unrational from science in general, as mystical, emotional, or womanly.

This conception of subjugated knowledges is not without its problems. For example, in terms of the discourse system of the present U.S. military, it would seem obvious that the viewpoint of the warrior (as opposed to the technocrat, manager, or timeserver) represents a subjugated knowledge. Yet the idea of the warrior seems to include not only critics of managing war as a technoscience and as an antidemocratic policy (Byron, 1985; Donovan, 1970; Newell, 1986) but also others whose conception of themselves as warriors encourages them to violate the law in order to make war against communism or some other pure evil (Ehrenreich, 1987a; W. Kennedy, 1987).

Despite the complexity of discourse, it seems a useful approach, especially because it helps explain a crucial dynamic: how does something like war change? Foucault suggested several possibilities:

1. *Changes in the actual material knowledge of a technoscience*—for example, the development of probability theory led to important changes in medical discourse since it meant that epidemics and infection could be viewed in a new way.

2. *The influence of official or other organized power* (e.g., medical and other professional associations, the police behind the judiciary) can directly change the rules of a discourse system; Foucault demonstrates this by citing cases of mental illness, prisons, and medicine (1972, 1975, 1977, 1980).

3. *The insurrection of subjugated knowledges*—Foucault argued that it is through the "insurrection" and "reappearance" of these knowledges that criticism does its work on the essential problem, "changing the regime of the production of truth" (1980, p. 133).

Change doesn't always spring from within a power/knowledge system. It can also come from outside the defining rules of discourse—imported, incorporated, or catalyzed by relationships with other discourse systems or communities. There is a border around a system of power/knowledge, but it is permeable and shifting. There are also overlapping discourses. Some of them are veritable metadiscourses permeating a whole spectrum of more limited discourses, including other metadiscourses. War is certainly a metadiscourse, as basic elements of war are part of countless cultures over a vast sweep of time and space. Gender is another such metadiscourse, and one that is intimately connected to war. It may seem a digression to try explaining the gendering of war in the middle of an argument about a new type of "postmodern" war that itself is based in part on an examination of the role of computers, as weapons and metaphors, in battle. But it's not. The recent shifts in war's gender coding explain the development of postmodern war as well as anything does, and they are part and parcel of the phenomenon as a whole besides (Gray, 1992).

The Gendering of War

War is clearly a "man's thing," as one of my students once remarked. Yet, war would be impossible if women did not support it as mothers, wives, daughters, workers, and even soldiers, which Cynthia Enloe makes very clear in her book on women and war, *Does Khaki Become You?* (1983). But the role of women has certainly been subordinate to the part men play. Norman Dixon points out in his analysis of military incompetence that "it is in the nature of military organizations to recapitulate the psychodynamics of an authoritarian family group, one in which the paterfamilias can do no wrong" (1976, p. 218). Dixon (1976) has an illuminating chapter called "Anti-Effeminacy" on the enforcement of masculinity in the military. He recounts various British practices such as forbidding piano playing and denouncing art and cigarettes, as well as the shame of performing defensive assignments like convoy duty. Dixon lets Gen. Adna R. Chaffee introduce the chapter with his pithy remark, "Let war cease altogether and a nation will become effeminate" (p. 208).

There are several important aspects to the love some men have for war. First, there is pleasure. Even modern war has afforded men much pleasure. The World War II combat veteran and philosopher J. Glenn Gray has been one of the most honest commentators on this:

> What are these secret attractions of war, the ones that have persisted in the West despite revolutionary changes in the methods of warfare? I believe that they are: the delight in seeing, the delight in comradeship, the delight in destruction. Some fighters know one appeal and not the others, some experience all three, and some may, of course, feel other appeals that I do not know. These three had reality for me and I have found them also throughout the literature of war. (1959, p. 33)

William James, many years earlier, held a somewhat similar view: "The horror is the fascination. War is the *strong* life; it is life *in extremis.*" In order to do away with war, James argued, there had to be some way to fulfill its functions because in many ways war defined masculinity. Any replacement for war "must make new energies and hardihoods continue the manliness to which the military mind so faithfully clings" (1911, pp. 276, 287; emphasis in original).

But beyond war's direct pleasures there are indirect advantages. It certainly supplies a justification for privilege, especially of men over women. As such it also becomes a key element for many men (and women) in defining masculinity.

Since war is still an integral part of male identity it remains central to both positive and negative conceptions of masculinity, despite its horrific elements, no matter how much its emotional importance is denied. In fact, this denial can lead to serious misperceptions, as James Fallows argues:

> Finally, arguments about defense that lose sight of the facts have been shaped by unexpressed emotion. Along with relations between the races and relations between the sexes, defense is one of the three areas in which public policy is most likely to be skewed by deep psychological forces. The subject under discussion—war—is life's most abhorrent activity, but also one that, through the millennia, has been, in many eyes, the ultimate manifestation of masculinity. All preparations for war are preparations to do what is universally not only condemned but also celebrated by its survivors and many onlookers. (1982, pp. 182–183)

Bernard Brodie, the influential theorist of naval and nuclear strategy, made a similar point:

> Human emotions, including and especially repressed emotions, make up a vitally important part of the reasons why men resort to war or, being in an obviously unprofitable war, find it so difficult to withdraw from its clutches. Emotions enormously affect perception as well as decisions and behavior, and they certainly affect the degree of rigidity we show about any of these. No doubt there is a good deal of aggression in the normal being, especially the male human being, and just as surely there is much repressed rage among various personalities who may rise to positions of influence and power. There can hardly be any doubt that these factors are involved in the genesis of wars and in accounting for the intensity and especially the persistence with which they are usually fought. (1973, p. 312)

These factors have been explored in part already, especially by J. Glenn Gray (1959), quoted above. But some of the more insightful observations come from feminist scholars.

One of the best is Carol Cohn. After spending a year as a visiting scholar at MIT's prestigious Center for International Studies, she wrote a detailed account of the rhetoric of nuclear strategic discourse (1987). Her analysis shows how this discourse often revolves not around people and their needs but weapons (the subject in both senses), how metaphors far removed from the reality of nuclear weapons are used (sexual, the farm, food, friendship), and how, among themselves, the nuclear analysts often slip ironically into religious symbolism, even calling themselves "the nuclear priesthood" (p. 21). She claims that on this level their use of language reveals

> a whole series of culturally grounded and culturally acceptable mechanisms that make it possible to "think about the unthinkable," to work in

institutions that foster the proliferation of nuclear weapons, to plan mass incinerations of millions of human beings for a living. [They use] language that is abstract, sanitized, full of euphemisms; language that is sexy and fun to use; paradigms whose referent is weapons; imagery that domesticates and deflates the forces of mass destruction; imagery that reverses sentient and non-sentient matter, that conflates birth and death, destruction and creation.

This language is actually part of their appeal for legitimacy. It is a claim based on "technical expertise" and on the "disciplined purging of the emotional valences that might threaten their objectivity." But under the "smooth, shiny surface of their discourse" of "abstraction and technical jargon" she uncovered

strong currents of homoerotic excitement, heterosexual domination, the drive toward competency and mastery, the pleasures of membership in an elite and privileged group, of the ultimate importance and meaning of membership in the priesthood, and the thrilling power of becoming Death, shatterer of worlds. (pp. 41–43)

The psychodynamics of this turn are complex and incompletely understood. Cohn mentions two possibilities she postulates from the work of the feminist theorist Dorothy Dinnerstein (1977):

1. That men involved in war-making and life-threatening actions manage their ambivalence over this activity by externalizing the negative parts of their feelings onto a group or object that is held in low value, that lacks power, and that is traditionally marginalized, even "ignored or scorned"—women being a prime example
2. That technical and scientific creation projects sometimes represent attempts by men to "appropriate from women the power of life and death" (Cohn, 1987, pp. 17–19).

One doesn't have to accept these theories uncritically, and I do not, to be struck by the ways the language of the nuclear bomb makers supports Cohn's interpretation. That so many witnesses of atomic blasts—from Trinity, the very first one (at Alamogordo, NM, on July 16, 1945)—have felt themselves part of some great creation is difficult to explain away. The first bomb was "Oppenheimer's baby"; the hydrogen bomb was "Teller's baby." On seeing his baby explode J. Robert Oppenheimer commented that "It was as though we stood at the first day of creation." He then quoted the words of Shiva, the great multiarmed goddess-mother, from the *Bhagavad Gita*: "I am become death, destroyer of worlds."

William Laurence, a U.S. War Department historian, said on seeing the first Trinity test: "One felt as though he had been privileged to witness the

Birth of the World." Maj. Gen. Leslie R. Groves' official cable reporting success of the same test to Secretary of War Henry L. Stimson read, "Doctor has just returned most enthusiastic and confident that the little boy is as husky as his big brother."

Cohn has many more stories (duds were girls) and a subtle analysis of the gendering of the early bombs (pp. 19–24). Jane Caputi (1987) has similar examples. The bomb that was dropped on Bikini atoll was a real babe. It had a picture of Rita Hayworth on it, and it was named Gilda after a movie of hers (p. 156). Most startling of all, Oppenheimer was named "Father of the Year" by the American Baby Association for making the atomic bomb (p. 187).

In his work the psychohistorian Robert Jay Lifton remarks on a related dynamic—that the fear of nuclear weapons can become a love for them:

> Nuclear weapons alter and blur the boundaries of our psychological lives, of our symbolic space in ways crucial to our thought, feelings and actions. The most extreme state of contemporary deformation is a pattern which may best be called "nuclearism." By this term I mean to suggest the passionate embrace of nuclear weapons as a solution to our anxieties (especially our anxieties concerning the weapons themselves). (1970, pp. 26–27)

He connects images of the extinction of the human race with dreams/hopes for the human future. Lifton points out that Star Wars, with its intense futuristic expression, has a deeply nostalgic form, the return to a time before we were all threatened by nuclear extinction. It represents a general form of nuclearism he calls "technism," which he defines as "an absolute embrace of technology for warding off an ultimate threat to human existence." Technology not only becomes a shield for the humans but in many ways it seems headed toward "literally replacing human responsibility" (1987, p. 125).

Along with the redirection of fear to love/gratitude that Lifton calls nuclearism, he also postulates something analogous to Dinnerstein's "externalization" in the idea of "doubling." From his research with atomic bomb survivors, nuclear strategists, elite college students, and Nazi death camp doctors, Lipton has concluded that under extreme stress many people create psychic doubles. These "others" then go on to perform the tasks the "real" person did not want to, or could not, take responsibility for.

The irony is that while everyone lives within this system, the people most likely to create alternate personas are exactly the same people with the greatest chance of changing the way things are. Yet, in many ways they are the most trapped by the very webs of power and desire that they weave.

Some intriguing speculations of a general structure of this network of power and desire in these discourses can be found in the work of the Australian philosopher Zoë Sofia (1984). She studies what she terms the "sexo-semiotics" of technology. She agrees with Foucault that in the nineteenth century the general technologizing of power, accomplished by the inscription of power on bodies and of pleasures onto the body politic, also involved the creation of an "analytics of sexuality" that in some ways was the technologization of sexual discourse. Sofia points out that this was accompanied by a balancing sexualization and mythification of technology. Where Foucault shows how disciplining technologies produced categories of perversions, she looks to undisciplined technologies "proliferating embodiments of perversions."

She accepts Norman O. Brown's assertion (1959) that technology is invested with erotic and bodily energy by the labor process, leading to an alienation by the human that seeks to animate the unliving while it reacts with disgust to its own living body. She traces this not to the child's desire to emulate the father, as Brown does, but rather to masculine envy of female reproductive capacity—womb envy, as Cohn suggests as well.

Much of Sofia's evidence is in the form of analysis of science fiction literature and movies and is too elaborate to present here. What it points toward is a set of detailed predictions about the epistemophilia, alienation, and other psychological associates of technoscience that seem confirmed by the intense, barely hidden, emotional energy revealed in the military's commitment to computers.

Sofia suggests that

> The Big Science worldview emphasizes epistemophilia (obsessive quest for knowledge, especially of origins), upward displacement (the High—or extraterrestrial—of hi-tech), and half-lives (projection of body-life into machine-life and vice versa). (1984, p. 59)

Sofia sees many of these forces or "libidinal energies" as "displaced from body to machines," leading to man's preoccupation with "animating excrement" and "resurrecting dead matter"—in other words, creating artificial intelligences. Such creationism runs on very set tracks. It is rationalistic in a crude lockstep way that insists every element of life and of consciousness is either unimportant or translatable into numbers and their relations.

This vulgar rationality (and no doubt the emotions that fuel it) is a major justification for war and it can still confuse seasoned observers. Military historian Michael Howard concluded after a lifelong study of the subject that

> The conflicts between states which have usually led to war have normally arisen, not from any irrational and emotive drives, but from almost a superabundance of analytic rationality. . . . Men have fought during the

past two hundred years neither because they are aggressive nor because they are acquisitive animals, but because they are reasoning ones. . . . Wars begin by conscious and reasoned decisions based on the calculation made by *both* parties, that they can achieve more by going to war than by remaining at peace. (1984, pp. 14–15, 22; emphasis in original)

If this is truly so, we should start to worry that maybe something is wrong with rationality. But it is only the most shallow rationality that is used to justify wars, especially in the modern era. Susan Mansfield is much closer to the mark when she argues that wars are actually fought mainly for psychological reasons. In her view, war is

a human institution that satisfies deep-seated psychic needs (the infantile desire for revenge on powerful parents, the anxiety-based insatiability for goods and power, a paranoid sense of powerlessness, etc.) and as a ritual attempt to force nature and the divine (the environment) to conform to human will. (1982, p. 19)

Yet, neurotic as it may be today, war still has a tremendous grip on human culture. To understand why this is so we will have to cover a great deal of ground, beginning with a brief exploration of the battles of thousands of years ago.

The Roots of Modern War in Primitive and Ancient War

Out of the warlike peoples arose civilization, while the peaceful collectors and hunters were driven to the ends of the earth, where they are gradually being exterminated or absorbed, with only the dubious satisfaction of observing the nations which had wielded war so effectively to destroy them to become great, now victimized by their own instrument.
—*Quincy Wright (1964, p. 42)*

There are many different theories about the origins of war. Most use one or more of these explanatory stories:

- War as the first machine with men as the parts
- War as a power play of unemployed hunters led by priests
- War as men's equivalent to birth
- War as a way of life for barbarian nomad-warrior cultures
- War as cancer (warrior cultures taking over peaceful cultures)
- War as the health of the state
- War because of elite or mass misperceptions
- War as testosterone poisoning
- War because humans are aggressive animals

- War because human brains are an evolutionary dead end
- War for profits or prophets
- War as a ritual
- War as progress
- War as a self-perpetuating machine
- War as racism

All these versions of war have more than a little truth behind them. They differ because they represent different levels of explanation. Each story explains some aspects of war better than others. Which might be best for analyzing postmodern war?

One way to begin is by asking the first question of Carl von Clausewitz: *De quoi s'agit-il?* What is it for?

Wars by Neolithic peoples, often called primitive wars, served population and ritual needs. Early wars by city people and the early empires, so-called ancient wars, also seemed to revolve around economic and ritual needs, with notable exceptions where quite different cultures became embroiled in total wars of survival, either of people or of ideologies.

For the last 500 years war has been conceptualized as an extension of politics. In Clausewitz's famous phrase "war is politics by other means." Since World War II clearly wars of survival still occurred; but with the technoscience readily available to produce nuclear weapons, war as merely politics has become nonsensical (Brodie, 1973).

Yet, small wars burn fiercely all over the globe while noncombatants supply the arms and build for themselves the largest, most destructive arsenals in history. Of the many possible reasons for this tenacity on the part of war, only those that most effect recent history and the types of war it has produced today can be examined here. The two premodern types of war, ritual and ancient, each contributed differently to contemporary war.

Primitive, heroic, unorganized, ritual war may seem very distant from the video images of Desert Storm, but it still exerts a force in our culture, and not just in war. Sports are in some ways more like ritual war than war itself is now. Still, despite the many changes in war today it is still strongly shaped by its ritual, clearly patriarchal origins. For example, the special role of weapons in precivilized war has left a lasting imprint on battle ever since. The naming of weapons, the granting to them great magical powers, and the assumption that they fight somewhat autonomously are all ideas that have never died out, and today they are undergoing a remarkable resurgence.

To chart the intimate relationship between weapons and war is not to say that there is a necessary relationship between invention and war. New tools do not have to become new weapons. They become weapons only if that is what seems natural to the inventors. As John Nef remarks in his book *War and Human Progress*:

> A revolution in weapons was a necessary part of the revolution in tools only in the sense that the Western peoples had not managed to break away from the traditional attitude of civilized as well as of primitive peoples—the attitude that weapons are one indispensable kind of tool. (1963, p. 41)

Nef's study shows how the inventions of the Middle Ages and the early industrial revolution were not based on military necessity. However, it also showed quite clearly that "as long as war remains a part of human experience, peoples cannot change their methods of producing without also changing their methods of destroying" (p. 41).

The debate about the importance of weapons to victory rages still in military discourse.

> Tools or weapons, if only the right ones can be discovered,
> form 99 percent of victory. . . . Strategy, command,
> leadership, courage, discipline, supply, organization, and all
> the moral and physical paraphernalia of war are as nothing to
> a high superiority of weapons—at most they go to form the
> one percent which makes the whole possible.
> —Maj. Gen. J. F. C. Fuller (1943, p. 3)

General Fuller, an early proponent of mechanized warfare, is one of the purest examples of a technological determinist; Napoleon, who claimed, "The Moral is to the Material as three to one," obviously didn't agree (R. O'Connell, 1989, p. 179). This debate between those who believe that material reality determines everything and others who give human will the important role is one that haunts this story of computers and war. Both positions are correct, at times. It's sorting out the particulars that is difficult.

Ancient war was very different from the ritual war it supplanted. It was part of a very great change in human culture and consciousness, especially around the idea of mechanization. Lewis Mumford and others have said that armies were the first machines, the first soldiers being the working parts. Even if this isn't true, it seems clear that rules, order, and form are crucial elements of all organized war.

Arthur Ferrill comments on how war is defined by men using forms (line and column) in battle:

> At the risk of grotesque simplification let me suggest that "organized warfare" can best be defined with one word. The word is *formation*. . . .
> When warriors are put into the field in formation, when they work as a team under a commander or leader rather than as a band of leaderless

heroes, they have crossed the line (it has been called "the military horizon") from "primitive" to "true" or "organized" warfare. (1985, p. 11; emphasis in original)[1]

Whether its organizing principle is called *calculation*, as Sun Tzu termed it, or *mechanism*, as in Mumford's pejorative formula, or *formation*, as in Ferrill's definition, it is clear that ancient war was organized in a very different way from ritual war.

The break between ritual war and ancient war is greater than any since. Modern war developed out of ancient war and kept many of the same rules and most of the same metarules. Postmodern war is very closely related to modern war and ancient war as well. The details of their differences, however, occupy a major part of this book. But before they can be explored, the key elements of modern war have to be articulated.

Chapter Six

Modern War

The Rise of the Modern World System

> Modernization is best conceived not only as an *intra*-social process of economic development, but also as a world-historic *inter*-societal phenomenon.
> —*Theda Skocpol (1983, p. 70; emphasis in original)*

Modern war, marked by the incorporation of industrial and scientific techniques, dominated by the logic of total war, and structured by very specific definitions of logic and rationality, coevolved with modernism. Modernism[1] is a world system. It is a system of hegemony—militarily, economically, and politically for those countries with Western-style economies. To say "Western" is not to pretend that there is one West, *the* Occident, but rather to notice that there is a Western viewpoint that is hegemonic, not only in terms of the rest of the world but even in relationship to different discourses within Europe and the "neo-Europes."[2] Even as modernism has been produced through many different models of industrialization for the rest of the world (variously mixes of capitalist, communist, fascist, social democratic, liberal, conservative), the decisive element for most countries is their role as either colonies or neocolonies of the industrialized countries.

And what is the basis for this success? In a word, technoscience. First the technosciences of war and then the technosciences of industrialization. This expanding hegemony is detailed in three remarkable books: Geoffrey Parker's *The Military Revolution: Military Innovation and the Rise of the West, 1500–1800* (1988); Daniel Headrick's *The Tools of Empire: Technology and European Imperialism in the Nineteenth Century* (1981); and Kurt Mendelssohn's *The Secret of Western Domination: How Science Became the Key to Global Power, and What It Signifies for the Rest of the World* (1976). Together, they give a clear and comprehensive explanation of the victory of European colonialism that established the modern state system.

[109]

Parker's book concentrates on the first 300 years of modern war, during which European states came to control 35 percent of the world's land surface. Headrick's describes how in the 114 years between 1800 and 1914 that domination went up to 84 percent (Parker, 1988, p. 5).

It is no coincidence that the development of modern war coincided with the modern state system. The military innovations that mark modern war have proven crucial in maintaining Western military superiority around the world to this day.[3] In the sixteenth and seventeenth centuries it was cannons on ships and firearms in general. In the eighteenth and nineteenth centuries artillery and repeating guns of various types became central. In the twentieth century armor, airpower, and superweapons have been the clear margin of victory up until the French defeat at Dienbienphu in 1954. In all these eras industrial power, translated directly into large-scale manufacturing of weapons, has been the central organizing principle of the modern war system.

Much of modern war is a story of the Spanish blunderbuss against the clubs and spears of the Aztecs and Incas, British matchlocks against the swords and pikes of the Hindus, Royal Navy men-of-war versus Chinese junks and Arab dhows, U.S. Cavalry repeating rifles against Sioux and Arapaho, and machine guns by Gatling and Maxim–Vickers against Arab dervishes, Chinese Boxers, Mescalero Apaches, and massed Zulu spearmen.

Shrapnel was first used in Surinam, the percussion musket in China, and the minié ball in southern Africa. The dumdum bullet, for example, which expands on contact, tearing jagged holes in its victims, was outlawed by the 1868 St. Petersburg Declaration when all such exploding bullets were forbidden. However, the British said it didn't apply to dumdums because the bullets, made at the Dum Dum Arsenal, near Calcutta, India, were only used on natives. Surgeon Maj. Gen. J. B. Hamilton and Sir John Ardagh explained, "Civilized man is much more susceptible to injury than savages. . . . The savage, like the tiger, is not so impressionable, and will go on fighting even when desperately wounded" (R. O'Connell, 1989, p. 242).

The story of modern war is also a tale of the decline of moderation in war until it "perished in the fireball above Hiroshima," as Michael Glover puts it in his *The Velvet Glove: The Decline and Fall of Moderation in War* (1982, p. 15). Ironically enough, that very fireball marked the end, as well as the apotheosis, of total war. Limited war has returned to central stage since 1945, although moderation certainly has not.

Modern war was also a prime shaper of modern politics, but not just in the obvious ways through military coups, military policy, and militarism in industry and education. Military metaphors, values, and concepts also intimately shaped most modern revolutions from 1789 to 1968 (Virilio, 1986). Through practice and birth, most modern states were modern war states.

Along with technoscience and modern state politics, the other crucial aspect of modern war was economics. For hundreds of years the importance of industrial power and infrastructure in producing military power was clear to military men and politicians alike. Railroads, the American system of manufacturing, steel production, and many other key facets of the modern economy have been directly traced to military demands (M. R. Smith, 1985b). Recently, this relationship between economic strength and military power has become even clearer. Paul Kennedy's history of the modern empires, *The Rise and Fall of the Great Powers* (1987), shows that, at least in modern times, there has been a very direct relationship between the economic power of the dominant Western countries and their military power. While fluctuations occur, the two measures have never drifted very far from each other. Even nineteenth-century Prussia, that "army with a state," could only grow as powerful as its economy.

In the Middle Ages, European war was dominated by aristocrats, although peasants did most of the dying. But in the 1500s three important trends developed that mark the beginning of modern war:

1. Applying rationality to war instead of tradition
2. The development of administrative bureaucracies
3. The systematic application of science and technology

A number of historians have called Niccolò Machiavelli the first theorist of modern war and for good reason. At a time when war in Italy was often an almost bloodless ritual, orchestrated by tradition and performed by professionals, he advocated conscripting armies to fight bloody, decisive battles. For most intellectuals of his time war was beneath serious theoretical or practical consideration, and it was deemed of only minor political importance. But Machiavelli argued that war was an intimate part of civil life. In his introduction to his book, *The Art of War* (1990), he links the two in a formula that hundreds of years later Clausewitz could well accept:

> Many are now of the opinion that no two things are more discordant and incongruous than a civil and military life. But if we consider the nature of government, we shall find a very strict and intimate relation betwixt these two conditions; and that they are not only compatible and consistent with each other, but necessarily connected and united together. (quoted in Gilbert, 1943, p. 3)

While Machiavelli advocated conscription of free citizens with a stake in the war, it would be almost 300 years (1789) before such mass national armies would became dominant. Still, the expansion of the money economy in the sixteenth century led to the eclipse of the agricultural–feudal foundations of war in the Middle Ages. This opened the way for the city merchants and wealthy overlords to replace feudal military obligations with purchased professional soldiers, laying the foundations for permanent mass armies.

Machiavelli was especially perceptive in noting the return of total war aimed at the complete subjugation of the enemy. As he said of the skirmishing and negotiations of the small professional armies of Italy in the early 1500s: "That cannot be called war where men do not kill each other, cities are not sacked, nor territories laid waste" (pp. 13–14).

Perhaps most significantly, the historian Felix Gilbert points out, Machiavelli took advantage of the "political events of his time . . . the intellectual achievements of the Renaissance . . . the military innovations of the period" (the rise of infantry, new weapons), and the "crisis of political institutions and values which signified the end of the Middle Ages" to make a "radical reexamination" of military and political assumptions.

His first new assumption, to quote Gilbert, was that

> there were laws which reason could discover and through which events could be controlled. The view that laws ruled also over military events and determined them in their course was a fundamental assumption of Machiavelli's concern about military affairs; his military interests centered in the search for these laws. Thus he shared the belief of the Renaissance in man's reason and its optimism that, by the weapon of reason, man was able to conquer and destroy the realm of chance and luck in life. . . . When Machiavelli set the battle into the center of his military theories, he could do so because the uncertainty of the outcome which previous writers had feared did not terrify him; by forming a military organization in accordance with the laws which reason prescribed for it, it would be possible to reduce the influence of chance and to make sure of success. (p. 3)

Success is never sure in war. But along with Machiavelli, men began to believe that through reason one could almost be sure. An important part of this reasoning was the great improvement in the infrastructure of war: the bureaucracies and economies that raised and maintained the armies.

Michael Wolf (1969, p. 41) notes "the most important development in the organization of western European military power was the rise of the ministries of war and marine as bureaucratic organizations responsible for the conduct of warfare." Rupert Hall also emphasizes the importance of organization at the expense of scientific developments in the sixteenth and seventeenth centuries:

> Throughout the period with which I am concerned today military affairs were determined by three things: organizing ability, including what we today call logistics; basic craft skill; and courage. . . . They no more depended on scientific knowledge than they did on the relative size of the combatants' populations or their real economic strength. (1969, p. 4)

But Hall and Wolf are drawing hard lines between crafts and science and logistical bureaucracies and industry where none really appear. Science was nurtured by many forces, among the most important are the ones Hall emphasizes. In turn, science changed the crafts, the bureaucracies, and logistics. Besides the general matrix these social forces shared, the men of the new philosophy, now known as science, made many direct contributions to the changes in war that marked the birth of the modern age. Hall himself points out many specific instances. For example, navigation and artillery were strongly improved by scientific discoveries and approaches (Hall, 1969, pp. 5–6), and both the Royal Society and the Académie des Sciences were interested in ways of testing gunpowder (p. 10). In terms of fortifications, geometry was crucial. Bonaiuto Lorini said it was "essential, the very foundation of all our procedures" (quoted on p. 12).

Perhaps the clearest case in this period of science influencing war is artillery. All the authoritative artillerists of the time claimed science was necessary for artillery, and all used mathematical tables, even though they were not empirically developed but based on various theories of ballistics that have since been discredited, such as proportional symmetry and strictly rectilinear segments. Galileo produced the first correct tables in his *Discourses on Two New Sciences*, in 1638. "This work on ballistics was developed further by Galileo's pupil, Evangelista Torricelli, who generalized and completed the theory, after which it passed into general circulation" (pp. 20–21).

Wolf makes a similar point when he notes the improvements of cannon and mortars in the period 1670–1789 in terms of standardizing balls, reducing the charge of powder, better caissons, new types of harness, and better servicing equipment and procedures. All of these innovations were brought about by engineers and bureaucrats working for navy and war ministries. He also comments: "Perhaps of equal importance was the development of more effective weapons in the hands of the seventeenth century infantryman. By 1500–1550 firearms had completely displaced the cross- and longbows" (pp. 35–36). Over the next two hundred years there were many other improvements, including the fusil or flintlock, strap bayonets, better cartridges, and iron ramrods. These Wolf attributes to gunsmiths and sportsmen, not scientists.[4]

John Nef, on the other hand, argues clearly that "behind the development of artillery and small firearms at the end of the fifteenth century was [the] movement of scientific inquiry" (1963, p. 44). He also emphasizes that some of the earliest scientists were military men, who made their discoveries in battle. The leading engineer of his day, Pedro Navarro, who invented explosive mines and floating batteries, is a case in point. Nef concludes that

The instruments for inflicting pain and destruction were not prepared in laboratories as in our mechanical age; they were often improved with the enemy before the inventor's eyes; the mingled physical horror and perverse pleasure entailed in their use was more obvious. (p. 44)

Still, individuals and their discoveries only played a part in the great social changes of the time. "The importance of gunpowder," explains Felix Gilbert, was only in the context of three more general developments:

> first, the rise of the money economy; second, the attempts of the feudal overlord to free himself from dependence on his vassals and to establish a reliable foundation of power; and third, the trend toward experimentation in military organization resulting from the decline of feudalism. (1943, p. 6)

Over the next 300 years weapons continued to improve, the money economy to spread, experimentation in the military to increase, and the aristocracy to decline. The French Revolution dealt the final blow to the aristocratic hegemony in war. It is no accident that modern war was first labeled such then.[5] Four years later the French Republic called for total war. Bernard Brodie quotes from a decree of the National Convention from 1793:

> All Frenchmen are permanently requisitioned for service in the armies. The young men shall fight; the married men shall forge weapons and transport supplies; the women will make tents and clothes and serve in hospitals; the children will make up old linen into lint; the old men will have themselves carried into the public squares and rouse the courage of the fighting men, to preach hatred against kings and the unity of the Republic. (1973, p. 253)

Science was certainly mobilized at this point. The great mathematicians Gaspard Monge and Lazare Carnot, became, respectively, the ministers of Navy and War. According to John Nef:

> Their application of the new geometrical knowledge to the actual conduct of war, together with their application of scientific knowledge generally to the problems of supplying adequate munitions, were factors of importance in the shift in Europe from wars of position to wars of movement. (1963, p. 320)

Nef notes a large number of important scientists were involved in war work, especially mathematicians. The efficiency of artillery, from manufacture to deployment, increased enormously. Their work also began to spread the idea that battles could be won "by calculations in the relative calm surrounding desks" through logistics and ingenuity, like a "game of chess." But, although, the "pawns were real men," the bureaucrats had growing difficulty in distinguishing "actual from toy soldiers" (p. 322)—a difficulty that has yet to be really overcome.

Between 1814 and the Crimean War there were no major European wars. Still, the weapons were improved. In 1827 the needle gun was invented. In

1829, cable/telegraph networks were established. In the 1840s rifled artillery was perfected. In 1848 the Prussians moved a complete army corps by train in an exercise. In 1850 Morse code was devised (Brodie, 1973, p. 12).

During The Crimean War, 1853–1856, torpedoes, trench war, the war telegraph, steam driven floating batteries, and females nurses were all introduced. The telegraph even allowed distant officials in London and Paris to interfere in tactical operations "hour by hour in order to save losses and political credibility" (p. 17). But it was in the United States that modern war first reached its full development.

The American Way of War

America is one of the most warlike nations on earth.
—*Billy Mitchell* (*quoted in Franklin, 1988, p. 99*)

Americans have always looked to science for their answers, in war as in everything else.
—*Thomas and Barry* (*1991, p. 39*)

Geoffrey Perrat, in his history of the United States, *A Country Made by War*, traces the growth of the twentieth century's dominant world power in terms of the constant wars and militarizations that have made it possible. Now it is true that this image doesn't fit the self-conception of the average U.S. citizen, but the history is clear. Just consider how much of U.S. territory was won through war or how many presidents were war heroes.

Of course, the very founding of the Republic was through war, and a very interesting war at that. It marked the first successful anti-colonial revolution, as well as the first victory of an unconventional people's army over an established modern army. Is it any wonder that the British band played "The World Turned Upside Down" at their Yorktown surrender?

Most of the U.S. wars have been colonial. The most common were the hundreds of Indian battles and massacres whose end results were often genocidal, although in the early years the colonists were sometimes hard pressed (Steele, 1996). These wars were very important, despite their neglect by most military and social historians. In the Americas, in Africa, in Asia, and in Australia, the Europeans and the emigrant Europeans waged a long series of conflicts, often trying out new weapons and technologies.

In the United States advanced signaling, repeating rifles, and machine guns were all used extensively in various Indian wars. In most cases the strategy of total war was applied.

The Mexican–American War of 1846–1848 was a break from the Indian campaigns. The U.S. victory owed something to the growing industrial might of the country, but even more it was due to the skilled professional army that stormed Mexico. It was a war of volley fire and sharp battles. It wasn't a grinding conflict of attrition pitting one country's industrial muscle and human blood against another. That kind of war was still a dozen years away.

It was the U.S. Civil War, often termed the first industrial war, that ushered in the last phase of modern war. The importance of that bloody conflict (notable for the introduction of the Gatling machine gun, the metal warship, and the use of railroads and telegraphs)[6] cannot be overestimated.

Grady McWhiney and Perry Jamieson show in their book, Attack and Die, that the widespread adoption of the rifle in the 1850s made the aggressive frontal assaults that had won the Mexican War for the United States little more than mass suicide in the Civil War. The increased numbers and effectiveness of artillery didn't help matters either. Still, despite numerous mass slaughters of storming troops, very few Civil War commanders were able to adapt their tactics to the new realities of powerful rifles and massed artillery. The Union general Daniel Hill, who led his division in one of the many grandly heroic and bloody assaults, remembered later, "It was not war—it was murder" (quoted in McWhiney and Jamieson, 1982, p. 4).

One Civil War general who almost always avoided useless head-on assaults was William Tecumseh Sherman. He also played a central role in the spread of total war into the modern world. His famous March to the Sea was one of the first, and certainly the most famous, campaigns during a full Western war aimed directly at civilian morale and economic warmaking potential. Other U.S. generals (especially Ulysses S. Grant), and the Confederate cavalry raiders into the North (notably under Nathan B. Forrest, John H. Morgan, and William Quantrill), also practiced mass appropriations and other terror tactics, such as burning buildings from which shots had been fired. But Sherman made it the basis for a campaign and even allowed several whole towns to be destroyed by not tightly controlling his troops. He also, on at least one occasion, seized a whole factory full of textile workers and shipped over 400 of them to the North to make uniforms for the Union Army (Reston, 1984, p. 30).

James Reston, Jr., has written an insightful book, Sherman's March and Vietnam, that shows how Sherman's "counterinsurgency" policies led quite logically to the U.S. policy in Vietnam, which mirrored the Union approach during the Civil War (1984). Attrition, diplomacy, technology, and economic war (in that it aimed at economic power and civilian working morale) were the main pillars of the U.S. government's strategy in both cases. The difference was that the predominant Confederate strategy of aggressive main-force battles was quite unlike the "people's war" of the Vietnamese. The Vietnamese and the Confederates did share a realization that their best chance of victory was political, and it would increase as the conflict dragged on. The Confederates just couldn't last.

The United States's use of total war predates the Civil War by almost a hundred years, Reston argues. He notes that Sherman wrote about the rebellion of the Seminoles in language that mimics U.S. generals writing about the Vietnam War 120 years later (p. 90). The prolonged Seminole War in Florida was won in the only way the Vietnamese War could have been won—genocide. But Reston doesn't put the Vietnam War into the context of the 200 years of extermination wars against the Indians that preceded the Seminole War, especially in New England by the English colonists and in the mid-Atlantic region by the Dutch and English.

Reston also discusses in detail Sherman's later career as an Indian fighter. Before he took command of the war against the Indians in 1870 Sherman stated what his policy would be toward Indian resistance: "If I were in command, I would act with vindictive earnestness against the Sioux, even to the extermination of men, women, and children" (quoted in Reston, 1984, p. 90). Later, under his command, U.S. troops killed almost 200 Piegan Indians, a quarter of them women and children, while suffering only one casualty of their own. Even *The New York Times* termed it a massacre. Sherman defended the "battle" by referring to the shelling of Vicksburg and Atlanta, where women and children were also killed "out of military necessity."

At the beginning of the twentieth century, this strong U.S. tradition of war became linked completely to the application of technoscience to battle. The Spanish–American War, in which the U.S. Navy destroyed the Spanish fleets with hardly a casualty, was one example of how potent this combination could be. Even more impressive, in retrospect, was the much more difficult and bloody conquest of the Philippines and the numerous successful invasions of Latin American countries by the U.S. Marines, who used machine guns, artillery, telegraphs, and even airplanes to gain an advantage.

Before the Civil War the technological requirements of civilians and soldiers were not that different. Even West Point was basically more of an engineering school than a military one. Science wasn't really mobilized until World War I when the National Research Council was set up. Melvin Kranzberg points out that even though the National Academy of Sciences was created during the Civil War to help the military, in the first 50 years the War Department only asked for five studies.[7]

But in the realm of management, scientific and otherwise, and formal systems of manufacturing, the relationship between the U.S. military and the civilian sector was very strong, even intimate. The military itself is a formal system, although few call it logical. The rigid hierarchies in most armies, their insignias, uniforms, and the early historical codifications and written expert systems, as far back as Sun Tzu's *The Art of War*, show the strong attraction between war and systematic rationality. The military has always had a burning desire for rules and orders, all the better to meet the unruliness and disorder of battle. Through history most armies have depended on automatic obedience (usually at the threat of instant death) from its soldiers. The men had to be parts that marched together, charged together, fired together, and died together. When discipline failed and the formation was lost, defeat and massacre followed. But in the face of the growing technical aspects of battle, and to make possible continual innovation, traditional military discipline has become scientific and more businesslike. Leadership has become management.

There is strong evidence that the American system of management, and Fordism later, were both shaped by military managing innovations. Merritt Roe Smith summarizes some of this research and notes:

> When one understands, for example, that Fordism traces its ancestry to the military arms industry of the nineteenth century, one begins to appreciate how deeply military-industrial rationality and centralization are implanted in American culture. The history of virtually every important metalworking industry in nineteenth-century America—machine tools, sewing machines, watches, typewriters, agricultural implements, bicycles, locomotives—reveals the pervasive influence of military management techniques. (1985a, p. 11)

Military and industrial management coevolved with the new technologies. Smith argues that "technological innovation entails managerial innovation" and therefore "technology and management are inextricably connected." He goes on to claim that this substantiates Lewis Mumford's view that the "army is in fact the ideal toward which a purely mechanical system of industry must tend" (Smith, pp. 10–11, quoting Mumford, 1934, p. 89). But it wasn't until the eve of World War I that the relationship was formalized.

The U.S. Army's official interest in the scientific management of itself can be traced back to 1909, when Frederick W. Taylor's principles were initially applied at the Watertown Arsenal. Taylor's development and codification of earlier military–industrial management techniques was a serious attempt to seize control of the shop floor from the workers. The weapons were time studies, standardizing machines, schedules, accounting and inventory controls, time–motion studies, and more workplace discipline.

Taylorism was met with a great deal of worker resistance, including bitter strikes in 1908 and 1911 at the Rock Island and Watertown Arsenals. But by 1918 one-third of the members of the Taylor Society were working in the Ordnance Department of the Army (Aitken, 1985). Worker's strikes and anti-Taylor legislation from prolabor members of Congress were swept away by the U.S. entry into World War I. Militarization consolidated Taylorism in the arsenals and in many other industries.

Taylorism shows clearly the connection between military and industrial discipline and marks how the growth in the power of war managers has not only turned soldiers more and more into workers who produce death and destruction but has also turned workers into industrial soldiers, producers of war matériel (men and machines).

David Noble's monumental history of industrial automation, *Forces of Production*, traces the history of scientific management, automated manufacturing and related systems through most of the twentieth century. He also describes the failure of this paradigm to successfully discipline the labor force any more than it rationalized the conditions of war:

> Taylor and his disciples tried to change the production process itself, in an effort to transfer skills from the hands of the machinist to the handbooks of management. Once this was done, they hoped, management would be in a position to prescribe the details of production tasks, through planning sheets and instruction cards, and thereafter simply supervise and discipline the humbled workers. It did not work out as well as they planned. No absolute science of metal cutting could be developed—there were simply too many stubborn variables to contend with. Methods engineers, time study men, and even the Army-trained Methods-Time-Measurement specialists who emerged during World War II, however much they changed the formal ways of doing things, never truly succeeded in wresting control over production from the work force. (1986b, p. 34)

There is "no absolute science" of war either, but it has been pursued none the less. The military's acceptance of scientific management spawned large-scale psychological testing of recruits for the first time, although it did nothing to lower the number of psychological casualties, which was its goal (Gabriel, 1987). The intelligence tests, on the other hand, produced interesting correlations and, in Merritt Roe Smith's words, "helped to stimulate general interest in the use of aptitude tests by managers as instruments for matching employee skills to the requirements of different jobs" (1985a, p. 14). What started as an attempt to weed out unfit soldiers became a tool for placing workers, which in turn the military adopted for assigning its own personnel.

By World War I the military and capitalist paradigms had become almost one and the same. American culture was militarized. In Kranzberg's view:

> The military . . . has become a major factor in our educational institutions and a prime force in industrial and economic life—and these largely because of the scientific and technological underpinnings of modern warfare. . . .
>
> Indeed, all three elements of our concern—science, technology, and warfare—form an integral part of modern American civilization. . . . The increasing importance of military institutions in modern American life thus depends in large part upon the fact that the military has become inextricably intertwined with science and technology. As these have become significant components of our national life and our national

security, the military too has gained a major role in American society. (1969, pp. 169–170)

Still, the full blossoming of modern total technological war did not take place until the first of this century's world wars, the "war to end all wars" in H. G. Wells mistaken epigram—named by the generation that survived it: the Great War.

The Great War

I don't know what is to be done, this isn't war.
—*Field Marshal Lord Kitchener (quoted in Ekstein, 1989, p. 165)*

How can the horror of the Great War be communicated? Poet-veterans, moviemakers, and historians have all tried. All wars have their horrible moments, but World War I seemed to mark a turning point. Any glory, any heroism, any individualism, any chance of the soldier controlling his own fate were stripped away. As many have said, men no longer made war, war was made on men, and the killing ground was called no-man's-land.

Modris Ekstein quotes a story from a British officer's diary to show what war had become. It seems that rotting German corpses were making the British trenches uninhabitable. So volunteers were sent out into no-man's-land to burn the bodies. As the officer puts it, "many gallant deeds are performed." One brave man in particular manages to burn three corpses before he is shot dead. The diarist comments, "cold blooded pluck." Ekstein asks, "How long would it be before men sensed the horrible ironies of a world in which gallantry was called upon to fight corpses, in which the living died trying to destroy the already dead?" (1989, p. 221, quoting the diary of Brigadier P. Mortimer).

But it wasn't just the erasure of courage by the incredibly powerful killing technologies that made the war so terrible. There were conscious decisions by both sides, starting with the Germans, to attack civilians. As they did in the Franco–Prussian War (and many colonial encounters), the Germans shelled civilian neighborhoods and executed hostages, including women, children, and the elderly, almost from the start. They burned the library at Louvain, which was over 500 years old and had a quarter of a million volumes, many irreplaceable. They bombed the cathedrals at Rheims, Albert, and Paris (pp. 156–157). Later in the war they launched unrestricted submarine warfare. The Allies soon responded with bombings and shellings of their own, but they refrained from taking or executing civilian hostages, not that some didn't advocate it. The Right Reverend A. F. Winnington-Ingram, Bishop of London, proclaimed, "Kill Germans! Kill them! . . . kill the good as well as the bad . . . kill the young men as well as the old . . . kill those who have shown kindness to our wounded . . . " (p. 236).

Ekstein gives special credit to the Germans for accelerating the full flowering of total war. It's roots were in the total militarization of Germany, according to Ekstein, because for the Germans "all political questions, all economic questions, all cultural questions, were in the end military questions" (p. 146). Once the war stalemated, the strategy of bloody, industrial, attrition was followed. But its justification was almost unconscious, based as it was on the militarization of all problems.

> Now, attrition was to be merely an offshoot of such thinking. It could not have grown had there not been a consistent buildup toward "totality." This called for the breakdown of the distinction between soldiers and civilians and the rejection of accepted morality in warfare. The treatment of civilians in Belgium by the occupying German forces and the reliance on new methods of warfare—especially the use of gas and inventions such as flamethrowers, and the introduction of unrestricted submarine warfare— were the most important steps, until attrition, in the advent of total war. (p. 147)

Along with this flowering of total war, Ekstein marks out several other crucial aspects of World War I. First in importance was it's industrial, middle-class character. It was, he points out, "the first middle-class war in history":

> It is therefore hardly surprising that the values of this middle class should have become the dominant values of the war, determining not only the behavior of individual soldiers but the whole organization and even strategy and tactics of the war. It's very extent—it was of course called the Great War—was a reflection of the nineteenth-century middle-class preoccupation with growth, gain, achievement, and size. Machines, empires, armies, bureaucracies, bridges, ships, all grew in size in the nineteenth century, this maximalist century; and dreadnought and Big Bertha were the telling names Europeans applied to their most awesome weapons of the eve of the war, this maximalist war. (pp. 177–178)

The soldiers on the front lines recognized that the war was like work in an abominable factory. César Méléra said revealingly at Verdun that it was "the bankruptcy of war, the bankruptcy of the art of war; the factory is killing art." Benjamin Crémieux, another combat veteran, commented, "The worst horror of this war, was that the men who made it were able to do so with the same conscience as any other work" (pp. 184–185).

But this work of war had unexpected results. One of the great victories of modernism was conquering nature with railroads, dams, cities, fast ships, planes, scientific farming, zoos, and museums. In World War I this domination of nature became insane, or so it seemed to the men who were there. Paul Nash, wounded at Ypres and then returned to the front as a war artist, describes the devastation:

> Sunset and sunrise are blasphemous. . . . Only the black rain out of the bruised and swollen clouds . . . is fit atmosphere in such a land. The rain drives on, the stinking mud becomes more evilly yellow, the shell-holes fill up with green-white water, the roads and tracks are covered in inches of slime, the black dying trees ooze and sweat and the shells never cease. . . . They plunge into the grave which is this land. . . . It is unspeakable, godless, hopeless. (quoted in Stallworthy, 1984, p. 275)

Other wars had destroyed significant portions of nature, but the Great War cut a gigantic putrid wound across the heart of Europe from the Alps to the sea. One pilot called it "that sinister brown belt, a strip of murdered nature. It seems to belong to another world. Every sign of humanity has been swept away" (Hynes, 1991, p. 21). Unsurprisingly, for many people the belief in modernism, in progress itself, was fatally wounded. Ekstein notes that before the war modernism was "a culture of hope, a vision of synthesis." After the war it was "a culture of nightmare and denial." He goes on to quote Robert Graves, a veteran who awoke after one wounding to find himself in a morgue presumed dead, that the war provoked an "inward scream" and "the duty to run mad." For Ekstein, "The Great War was to be the axis on which the modern world turned" (1989, p. 237); the turn was from life to death:

> The Great War was the psychological turning point, for Germany and for modernism as a whole. The urge to create and the urge to destroy changed places. The urge to destroy was intensified; the urge to create became increasingly abstract. In the end the abstractions turned to insanity and all that remained was destruction, Gotterdämmerung. (p. 329)

And Gotterdämmerung came two decades later, at least at Dachau, Auschwitz, Treblinka, Hamburg, Dresden, Tokyo, Hiroshima, and Nagasaki. The military historian Robert O'Connell reaches a similar conclusion but puts special emphasis on how military technology, which had given Europe world dominance, had turned on the Europeans themselves:

> The Great War had a profoundly lasting and deleterious effect on Western man's view of himself and his civilization which cannot be explained solely in political terms. Rather . . . at the root of this crisis of morale was a sudden awareness, engendered primarily by the stalemate on the western front, that military power, when applied, had grown uncontrollable, and that this was directly attributable to weapons technology. This judgment has not changed essentially to this day. Yet the abruptness of this occurrence,

the wholesale discrediting of a factor largely responsible for three hundred years of political transcendence in a mere four years of war, was such a shock and raised such profound questions about the basic directions of Western civilization that it created a crisis of the spirit unparalleled in modern times. (1989, Vol. 2, p. 242)

This crisis includes the very definitions of humans and machines. Klaus Theweleit, in *Male Fantasies*, his study of World War I German veterans, noticed a deeply erotic and ambivalent relationship between the soldiers and mechanization. "The new man," as these warriors described him, was "a man whose physique had been mechanized, his psyche eliminated—or in part displaced into his body armor" (1989, Vol. 2, p. 162).

The prowar writings of the veteran Ernst Jünger, for example, often focus on an imaginary man whose "instinctual energies have been smoothly and frictionlessly transformed into functions of his steel body." Theweleit sees a clear "tendency toward the utopia of the body–machine":

> In the body–machine the interior of the man is dominated and transformed in the same way as are the components of the macromachine of the troop. For Jünger, then, the fascination of the machine apparently lies in its capacity to show how a man might "live" (move, kill, give expression) without emotion. Each and every feeling is tightly locked in steel armor. (p. 159)

Jünger affirms, "Yes, the machine is beautiful: its beauty is self-evident to anyone who loves life in all its fullness and power." In other words, the machine is alive. Theweleit draws links between "The Soldierly Body, the Technological Machine, and the Fascist Aesthetic" in a chapter of that name. He quotes Jünger, who said, "we must imbue the machine with our own inner qualities" in turn the "machine . . . should provide us with a higher and deeper satisfaction." In the end, the "machine" takes over from the body (p. 197). This self-mechanization performs a crucial function, a pleasurable function, for the soldier males—it allows them the release of killing and risking death. In Theweleit's words: "The crucial impulse behind the regeneration of the machine seems to be its desire for release—and release is achieved when the totality-machine and its components explode in battle" (p. 155). In this profane traffic between war, humans, and machines, in this dehumanization and mechanization, the psychological reality of cyborgs may first have been born. Theweleit's summary is all too clear:

> At the same time, the unity and simplicity of the object-producing machine is dissolved; the machine becomes an expressive multiplicity of semi-human aesthetic forms. Thus the human being becomes an imperfect machine, and the machine an imperfect human being, neither any longer capable of producing, only of expressing and propagating the horrors they

have suffered. Perversely distorted, both now become destroyers; and real human beings, and real machines, are the victims of their mutual inversion. The expression-machine airplane drops bombs on production machines, as the mechanized bodies of soldier males annihilate bodies of flesh and blood. The libido of such men is mechanized and their flesh is dehumanized through mechanization. (p. 199)

This was not the experience of everyone in the Great War, to be sure. But it was the experience of enough western front poets, enough protofascist "front-line fighters" such as Adolf Hitler, enough men and women in all, that the idea of war out of control, whether it was a good or bad thing, entered deeply into the Western mass consciousness.

Even as World War I ran out of control, turning battle into an extended killing machine and planting the seeds of an even greater, more horrible war, other forms of control were brought into being. The modern state expanded in many directions through increased taxation, the regimentation of the economy and labor, and the institutionalization of everything from passports to art patronage.

In the United Kingdom and the United States the first truly systematic attempt to control war correspondents and manage the news were implemented (Knightley, 1975). At the same time, in a related development, this war to end war marked the fantastic growth in the size and power of the U.S. political intelligence system, which has continued to gain in influence and importance to this day (Donner, 1980).

In the United States science itself came under more direct government management in the name of war. For the first time, Army and Navy projects operating under secrecy found their way into the laboratories of the nation's colleges, at over 40 campuses. Many disciplines, especially physics and chemistry, were militarized. The war "forced science to the front," in the words of George Ellery Hale, one science manager (quoted in Kevles, 1987, p. 138).

At the front, science did its best to make war worse. Professor Fritz Haber, the pioneer of gas warfare whose mistaken judgments convinced the German General Staff to use gas, said on receiving the Noble Prize for Chemistry in 1919: "In no future war will the military be able to ignore poison gas. It is a higher form of killing" (quoted in Kevles, 1987, p. 137).

Chemists and chemical engineers were the most active of World War I scientists, so some have called it a chemist's war. Their most famous products were the poison gases, but they also produced a new generation of more effective explosives for guns, mines, grenades, bombs, torpedoes, and artillery shells (Harris and Paxman, 1982).

There were great advances in the science of ballistics as well, in improved mechanical calculators and theoretical advances. Several founders of computer science worked in these areas; the most famous, Norbert Wiener, calculated ordnance tables.

These technical innovations led to an incredible increase in the lethality of the battlefield through improvements in artillery and automatic weapons. It is under these conditions that a problem that perplexes military planners to this day first occurred—the collapse of whole armies under the stress of the conditions and heavy casualties of technological war. John Keegan lists the major instances:

> A point was reached by every army at which either a majority or a disabling minority refused to go on. This point was reached by the French army in May 1917, when "collective indiscipline" occurred in 54 of 100 divisions on the Western Front; in the Russian Army in July 1917, when it failed to resist the German counter-attack consequent on the collapse of the "Kerensky Offensive"; in the Italian army in November 1917, when the Second Army disintegrated under German-Austrian attack at Caporetto. In March 1918, the British Fifth Army collapsed, as much morally as physically, and in October the German army in the west signified to its officers its unwillingness to continue fighting. (1976, p. 276)

Many of the horrible casualties these men rebelled against were caused by the military command's faith in the warrior spirit, despite the reality of machine gun bullets. The French Army, for example, stressed the attack above all, to the death of many thousands of French soldiers (pp. 69–72). German, British, Russian, Australian, and Italian units were often thrown to destruction in World War I under the mistaken impression that force of will could directly overcome force of fire.

But the reverse certainly seems clear. Force of fire does overcome directly the force of will. Eric Leed put it plainly:

> The sheer scale of technologically administered violence seemed to force the regression of combatants to forms of thought and action that were magical, irrational and mystical. . . . Magic is an appropriate resort in situations where the basis of survival could not be guaranteed by any available technology. (quoted in Holmes, 1986, p. 238)

It was often the machine gun, that most industrial of weapons, that reduced the soldiers to primitives. The machine gun was just over five decades old by 1914, but few military men realized that it would become one of the dominant weapons of the next 50 years. Invented by Americans, but little used in the Civil War, the machine gun was first widely deployed as a colonial weapon. In numerous battles in Asia and Africa the machine gun saved the day for the empire. At the battle of Omdurman in the Sudan six Maxim guns killed thousands of dervishes. G. W. Stevens described the scene:

> It was not a battle, but an execution. . . . The bodies were not in heaps . . . but . . . spread evenly over acres and acres. Some lay very composedly with

their slippers placed under their heads for a last pillow; some knelt, cut short in the middle of a last prayer. Others were torn apart. (quoted in R. O'Connell, 1989, p. 233)

Often the guns were owned and operated by private companies, taking the political control that preceded their economic restructuring of non-European countries. For many colonialists the machine gun proved the superiority of Europeans. John Ellis notes ironically in his brilliant book, *The Social History of the Machine Gun*:

> The Europeans had superior weapons because they were the superior race. With regard to the machine gun, for example, one writer assured his readers that "the tide of invention which has . . . developed the 'infernal machine' of Fiechi into the mitrailleur [sic] and Gatling Battery of our own day—this stream took its rise in the God-like quality of reason." Thus when the Europeans opened their bloody dialogue with the tribes of Africa it was only natural that they should make them see reason through the ineluctable logic of automatic fire.[8]

In one action between employees of the German East Africa Company and Hehe tribesmen, two men and two machine guns killed roughly a thousand native combatants. In Tibet one British machine gunner became ill at the slaughter. His commander told him to think of his targets as game. In Rhodesia a tribesman asked, who are "the naked Matabele to stand against these guns?" And in Nigeria, another tribesman remarked, "War now be no war. I savvy Maxim gun kill Fulani five hundred yards, eight hundred yards far away. . . . It be no blackman . . . fight, it be white man one-side war" (quoted in R. O'Connell, 1989, pp. 233–234).

The United States also found machine guns useful. They were quite important during the Spanish–American War, notably at the charge up San Juan Hill and during the pacification of the Philippines. They also proved effective in the suppression of the Boxer Rebellion. Yet, despite these military successes, neither the European nor the American armies that used the machine gun in colonial wars developed a real understanding of what it would mean when two modern armies met. David Armstrong, author of a historical study of the U.S. Army and machine guns, concludes:

> Creation of a coherent body of tactical doctrine was not a topic of major interest in the higher echelons of the army for much of the period before World War I. Like the majority of their European counterparts, most American soldiers did not understand the extent of the changes in warfare that had occurred as the result of fifty years of rapid technological progress; consequently they failed to realize how outdated their tactical concepts

actually were. Doctrine was no longer a set of relatively simple rules that prescribed how men and weapons were to be maneuvered on the field of battle; it was, instead, a complex intellectual framework that enabled soldiers to conduct the intricate operations required in an arena that was increasingly dominated by the machines and techniques of modern industry. (1982, p. 212)

So when the Great War came there was great confusion. Gen. Douglas Haig claimed that the machine gun would lead to the return of cavalry (Ellis, 1973, p. 130). In the trenches the horrific effectiveness of machine guns led to a number of different reactions. Some soldiers even idolized the machine gunners of the enemy who killed so well.

Among the machine gunners themselves a great dehumanization took place. Ellis quotes extensively from the experiences of one British officer, Lt. Col. G. S. Hutchison, who later wrote a history of the Machine Gun Corps. After most of his company was wiped out during the second battle of the Somme he got hold of a machine gun and caught a group of Germans silhouetted against the skyline. "I fired at them and watched them fall, chuckling with joy at the technical efficiency of the machine." Later in the battle he turned his gun upon a battery of German artillery "laughing loudly as I saw the loaders fall." Near the end of the war he helped crush the German offensive of 1918 and remembers machine gunning retreating Germans as "thrilling." On the same day he found a group of drunk stragglers from his own army. At machine gun point he forced them to attack the Germans. "They perished to the man" (pp. 144–145), he remarks cheerfully. What kind of man is this?

Barbara Ehrenreich recognizes him as a war-man. She points out that many of the men who started World War II

[did] not emerge on the plain of history fresh from the pre-Oedipal nursery of primal emotions, but from the First World War. That war was a devastating experience not only for the men who lost, like these, but for those who "won." . . . In considering the so-far unending history made by men of the warrior caste, it may be helpful to recall that it is not only that men make wars, but that wars make men. For the warrior caste, war is not only death production, but a means of *reproduction*; each war deforms the human spirit and guarantees that the survivors—or some among them— will remain warriors. (1987, p. xvi; emphasis in original)

As Ehrenreich says so well, World War I made the men who made World War II; so the one war certainly made the other, as wars have bred wars down through history. Still, World War II was unusual. It not only bred more wars, but it birthed a new type of war as well, postmodern war.

Chapter Seven

The Emergence of
Postmodern War: World War II

By means of technique man changes not only the method
by which he thinks *but the content of what he thinks about.*
—William Blanchard *(quoted in Sherry, 1987,
p. 248; emphasis in original)*

World War II: The Last Modern War

World War II was unique, not just quantitatively, but in technological quality
as well. The strategic bombing technology (planes, bombs, organization) that
manufactured the firestorms of Dresden and Tokyo, as destructive as Hiroshima and Nagasaki, is one example. But most significant was the development of nuclear weapons and computers. The atomic bomb changed the
parameters of lethality forever. Computers not only made the A-bomb
industry possible, they also changed ballistics and logistics and have made
over command, control, communications, and intelligence (C^3I). In some
aspects World War II was a postmodern war, especially as concerns these two
weapons and the whole system of strategic bombing. But fundamentally
World War II was a modern war because it continued modern war's quest for
totality, up to and including Hiroshima.

It is with good reason that World War II is often called the physicists'
war, for physics made the total weapon, atomic bombs, possible. Yet, as valid
as that label is, it disguises somewhat the pervasive role of formal logical
systems and other aspects of technoscience. The scientific and bureaucratic
management of soldiers, the press, domestic intelligence, and scientists,
which got such a good start in World War I, became full-blown psychological
testing, a gigantic propaganda apparatus, an international intelligence bureaucracy, and a vast network of secret military research labs in World War
II. War was mechanized, not just with blitzkriegs, so dependent on radios,
trucks, tanks, and airplanes, but in its very organization.

In many cases it was mechanized thinking, instead of machines themselves, that came first. Consider scientific management and operations research. Both of these formal/logical systems are rule bound, explicitly defined, and involve a great deal of mathematical calculation. They were carried out originally on mechanical IBM punch card machines, then with the aid of electric calculators, and finally by electronic computers.

It was in the logistics of producing and moving vast numbers of soldiers and war matériel that most successes for formal logic were recorded. The human ability to control and manage large numbers of things grew tremendously, thanks to these formalizations of bureaucratic behavior. This management included the soldiers' psyches as well. The U.S. Army's Research Branch of the Special Services Division conducted a series of attitude surveys that, according to the historians of sociology Robert Lynd and John Madge, were supposed "to sort out and to control men for purposes not of their own willing." Many social scientists worried that the whole of sociology would become mobilized for solving the "managerial problems for industry and the military" (quoted in M. R. Smith, 1985a, p. 14; see also Buck, 1985).

The increasing management of soldiers was matched by official control of the press (Knightley, 1975), the creation of a large international intelligence system controlled by the United States (Donner, 1980, pp. 52–68; R. H. Smith, 1977), and by the military mobilization of American scientists to an unprecedented degree (Flamm, 1987, pp. 6–7; Kevles, 1987). Michael Sherry, in *Preparing for the Next War*, notes that military research went from $13 million in 1939 to $1.5 billion by 1944 (1977, p. 126). Many scientists who worked for the military were sure that civilian scientists would make war much more effective. As the science writer J. Crowther and the scientist R. Whiddington put it in *Science at War*, their semiofficial history of British science in World War II:

> The romantic conception of war is becoming out of date. It is not consonant with the systematic, rational, scientific kind of warfare which is evolving from the inter-penetration of war and science. . . . For the traditional romanticism of war is the contrary of the civilian scientific spirit, and it is therefore natural that when the scientist begins to join in the conduct of war, he enters into it as a civilian. That is why the directors of operational research are generally civilians, and one reason why war in the future will tend more and more to be conducted in a civilian spirit. (1948, pp. 119–120)

It was to be a very unromantic war. And perhaps its least romantic aspect and the best example of mechanized thinking, wed to high technology and scientistic technological fanaticism, was the allies' strategic bombing policy. To understand the emergence of postmodern war in World War II, it is necessary to trace the development of strategic bombing from its beginnings in the dropping of hand grenades on tribal villages to the plans for intercontinental nuclear war, the *reductio ad absurdum* of total war.

The History of Strategic Bombing

The airplane appears, and a . . . new military philosophy [is] centered upon it. Then nuclear weapons and rocket vehicles come along, and these create wholly new conditions . . . including certainly possibilities of unprecedented evil resulting from war.
    —*Bernard Brodie (1973, p. 243)*

I still remember the effect I produced on a small group of Galla tribesmen massed around a man in black clothes. I dropped an aerial torpedo right in the center, and the group opened up just like a flowering rose.
    —*Mussolini's son Vittorio describing combat in Ethiopia (quoted in Virilio, 1990, p. 19)*

The first real theorist of strategic bombing was the Italian Guilio Douhet. In 1909 Wilbur Wright visited Italy with an airplane and Douhet was inspired with the idea of air power (Holley, 1988, p. 22). He felt that new technologies meant new types of war. "The form of any war . . . depends upon the technical means of war available" (quoted in Holley, 1953/1983, pp. 12–13). Even before World War I he was calling for strategic bombing, so he deserves some credit for the sad fact that it was Italians who first dropped bombs from planes onto people. In October 1911, during the Italian–Turkish War, they bombed Turkish troops and Arab tribesmen in Libya. The pilots were probably trained by the Wright brothers, whose aggressive advocacy of military air power had earned them the Italian Air Force's training contract in 1909 (Donnini, 1990, pp. 45–51). A year later the French Air Force used terror bombing to put down an anticolonial rebellion in Morocco. Targets included villages, markets, flocks of sheep, and fields of grain (Kennett, 1982, p. 15).
    During World War I the Italians ran one of the largest bombing campaigns but they were matched by the Germans and, later, the British. Strategic bombing was first tried on October 11, 1914, when a couple of German Taubes dropped 22 bombs on Paris, killing 3 citizens, wounding 19, and scratching Notre Dame (Ekstein, 1989, p. 158). Almost a year later, on September 8, 1915, Zeppelin L-15 of the German Army bombed a number of London neighborhoods causing 72 casualties, including 17 astounded drinkers at the Dolphin public house on Red Lion Street (Dyer, 1985, p. 84).
    Only a few thousand tons of bombs were dropped on strategic targets in World War I, an amount soon matched in various colonial bombing cam-

paigns by France and Britain. The French even developed a fighter-bomber for just such a role, *Type Coloniale*, while the British initiated in parts of the empire a system of air rule called "Control without Occupation." In 1922 Air Marshal Sir John Salmond became the first air officer to command a combined invasion force, when he took charge of all British units in Iraq. The British also waged an air campaign against the Pathan (Pashtun) people on India's North-West Frontier, which included the bombing of villages and reservoirs. U.S. Marines bombed Ocotal and other Nicaraguan towns in 1927 (Franklin, 1988, pp. 88–89, 97). But these were just colonial bombings and so drew little notice. For many, bombing "civilized" people was still criminal. The strategic bombing of Guernica and Barcelona by the Germans during the Spanish Civil War, for example, was widely condemned.

On the day World War II started, September 1, 1939, President Franklin D. Roosevelt even issued an "urgent appeal" that "under no circumstances" should civilians or unfortified cities be bombed from the air. He called such bombing "inhuman barbarism." The year before, Secretary of State Cordell Hull had said of the bombing of Barcelona, "No theory of war can justify such conduct." That same year the U.S. Senate had condemned the "inhuman bombing of civilian populations." Yet, while American politicians were denouncing these attacks, the U.S. Army Air Force was itself preparing for strategic bombing, developing the new planes, bombs, institutions, and doctrines that would make the United States the world's leading strategic bomber by 1945 (Franklin, 1988, p. 101; Millis, 1956, p. 241).

Before World War II and during the early phony war stage, the idea spread of strategic bombing as a knockout blow aimed at the will, and perhaps the economic heart, of the opponent. But as the bombing surveys found after the war, the Germans, the British, and the Americans all failed to achieve the predicted success with their strategic bombing despite the great effort expended in men and material. Bombing stiffens civilian morale and does not totally disrupt production. Although it has yet to work, and it means killing civilians, strategic bombing still has many believers. It has been tried often, recently in locales as different as Vietnam, Afghanistan, Chechnya, and Iraq.

Freeman Dyson, later a well-known physicist, worked for Bomber Command during World War II, and he noticed early on that not only was his work not effective but it was also immoral (1984, p. 60). Dyson remarks with approval that Winston Churchill blocked the building of a monument for Bomber Command, although it was the only command without one. Bernard Brodie (1946, 1973) and John Kenneth Galbraith (1969a, b, 1981) were involved in studying the effects of the strategic bombing by the United States and United Kingdom and have written about their views extensively. Albert Speer, from his different perspective, reached similar conclusions, as is detailed in his memoirs (1970).

But the appeal of strategic bombing is so strong that despite its failure to be a decisive weapon, its supporters refuse to stop thinking it is. Science fiction writers and military theorists who believed in strategic bombing had long felt that mere explosive bombs would not be destructive enough to win a war. They argued that poison gases and incendiaries would increase the power of the bombing and achieve the desired effect (Franklin, 1988). As the use of poison gas was restricted for various reasons, especially the fear of retaliation, it was left to incendiaries to vindicate strategic bombing. With the help of scientists, military aviators calculated that by using the right mix of incendiaries and explosives on the right target at the right time a unique, man-made, weather condition could be created: a firestorm. They were correct.

The first successful firestorm was made in Hamburg, Germany. Operation Gomorrah, four assaults in late July and early August of 1943 by an RAF force of 731 attacking bombers, started the firestorm when the air heated to 800 Celsius, creating a gigantic blast furnace effect over the city. Most of the victims asphyxiated or were burnt into *Bombenbrandschrumpfleishen* (incendiary-bomb-shrunken bodies). More than 40,000 Germans died. The Japanese were next.

The Burning of Japan

American policy is to expend machines rather than men.
—*Intelligence Officer, 20th Air Force*
(*quoted in Sherry, 1987, p. 192*)

The idea of burning Japan had long been popular in the United States. In an article in the mass circulation weekly *Liberty* from January 30, 1930, entitled "Are We Ready for War with Japan?" Billy Mitchell, commented that Japanese cities were "an ideal target for air operations." A horrific drawing of the bombing of Japanese civilians illustrates the first page of this article (Franklin, 1988, p. 70). In 1943 Walt Disney even made a full-length cartoon about it, *Victory Through Air Power*, which begins with a clip of Billy Mitchell giving a speech in favor of strategic bombing.

Disney was inspired by a book of the same title as the cartoon published in 1942 by the head of Republic Aviation Corp., the Russian emigré Alexander P. de Seversky. Seversky called for "a war of elimination" against Japan. Disney's animation was exuberant, showing armies of bombers burning acres of cities without any human suffering. The climax is the transformation of the bombers into an American eagle that then claws the Japanese octopus to death. In a review James Agee called it "gay dreams of holocaust." Bruce Franklin comments on the end of the film: "As the audience beholds the smoldering remains of this make-believe nation, it hears the swelling strains of 'America the Beautiful.' Then across the screen is emblazoned 'VICTORY THROUGH AIR POWER' " (p. 108).

But it took the U.S. Army Air Force and its scientific experts to make it happen. And the instrument was the 20th Air Force, commanded directly from Washington, D.C. by Gen. H. H. ("Hap") Arnold, and at the front, 21st Bomber Command, under Maj. Gen. Curtis LeMay.

The German firestorms, for others were created after Hamburg, were closely studied by the Americans. One expert was a young OSS (Office of Strategic Services) analyst, Charles Hitch. After a trip to England he recommended using incendiaries on Japan. In Washington, D.C., he managed the "physical vulnerability directorate" in the Pentagon and worked on several bombing studies of Japan. One of these, finished before the fire-bombing of Japan started, predicted half a million fatal casualties, almost 8 million people "dehoused," and a 70 percent destruction rate on the targets. "The attacks assumed in this study would effect a degree of destruction never before equaled," the report enthused. Despite this, in the conclusion it was admitted that "it is unlikely that output in any one important [warmaking industrial] category will be so reduced as substantially to affect front line strength." In other words, despite killing half a million people it would not hurt Japan's ability to fight at all. It was to be terror bombing plain and simple to break the will of the Japanese to fight.[1]

Other scientist-analysts went out of their way to push for the fire-bombing as well. Dr. R. H. Ewell wrote a memo to Vannevar Bush, Director of the Office of Scientific Research and Development, that called "incendiary bombing of Japanese cities" a "golden opportunity of strategic bombardment in this war—and possibly one of the outstanding opportunities in all history to do the greatest damage to the enemy for a minimum of effort." He went on to bitterly complain that the Air Staff and 20th Air Force planners were blocking firebombing.[2] He need not have worried.

General LeMay became convinced that the opportunity of burning down Japan was too great to pass up. Not that it would win the war, but rather because it would prove "the power of the strategic air arm." In a revealing Telecon memo he sent to Brig. Gen. Lauris Norstad during an argument over how hard they were working the air crews, he admitted as much.

Both my surgeon and my wing commanders are convinced that to require this rate [the "short-term" rate of 80-plus combat hours per month] for a six-month period might burn out my crews. . . . In choosing between long-term and short-term operating policy I am influenced by conviction that the present state of development of the air war against Japan presents the AAF for the first time with the opportunity of proving the power of the strategic air arm. . . . Though naturally reluctant to drive my force at an exorbitant rate, I believe that the opportunity is [*sic*] now at hand warrants extraordinary measures on the part of all sharing it.[3]

The incendiaries themselves were made by the Army Chemical Warfare Service, the National Defense Research Committee, and the petrochemical industry. Much of the experimental work was directed by the Harvard chemist Louis Fieser. Fieser was something of an enthusiast. He tried to teach bats to carry tiny incendiaries so that when dropped from bombers they would nest in attics and start many efficient little fires. He abandoned this idea after he burned up the theater, officer's club, and a general's car in tests at the Army airfield at Carlsbad, NM (Sherry, 1987, p. 226). But strategic bombing was not just a matter of technology; it was first of all a system, and humans were key components.

To understand this a closer look at one analyst's history will be helpful. This young man was a professor at Harvard Business School when the war began. He was an expert in bargaining theory, but his work for the 20th Air Force was calculating the various options and necessities of strategic bombing.

For example, in one of his reports in 1945, he points out that at the possible "loss rates" of earlier months, the "proposed activity rates" of the 20th Air Force could not be sustained. Note the language of economics, as well as of mathematics, that frames the analysis:

> At the loss rates sustained during 1 March to 18 March and with a tour of duty allowing a 70% chance for survival, the present aircraft and crew replacement planning factors will support the proposed activity rates. However, an increase in the loss rate to the levels of January and February would require an increase in the aircraft replacement requirements and either an increase in the crew replacement rate or a lengthening of the tour of duty.

He does advocate that the replacement rate be picked up, if possible, which was the recommendation accepted. One of the more interesting graphics in this report is called "Tables of Survival." He shows how at the loss rate of 1.1 percent per sortie, a 30-sortie tour would be a loss rate of 26 percent, "well under the AAF standard of 35 to 50%." In other words, the crews can expect a 1 in 4 chance of being shot down before they complete their tour of duty.[4]

In another report in the same month he calls these crews "inventory" and notes that their survival rate was 65 percent over the previous six months.[5] For this young Harvard professor, whose survival rate was always significantly above 65 percent, this was a scientific, even economic, calculation, with no morality or fuzziness at all about it.

And as he calculated his inventory of air crews versus their losses, he also advocated a switch from precision bombing to firebombing because of its greater "efficiency."[6] His voice, no doubt, did not yet carry much weight. For during World War II he was only a small part of the strategic bombing system. But later, Robert Strange McNamara would run a war of his own.

When McNamara was first commissioned a captain he began doing logistical analysis on the operations of B-17s (in England for four months) and of B-29s (in Kansas for six months). During this period, 1943, he met General LeMay. In April 1944 McNamara was sent to Calcutta, India, to work directly with the strategic bombing force. Six months later he was made a lieutenant colonel, and that October he returned to Washington, D.C., and the Office of Management (Trewhitt, 1971, pp. 36–40).

His military service was working for the Statistical Control Office with human computers and IBM electrical–mechanical machines to keep track of various key variables for Army Air Force projects. All of his most important postings were to help strategic bombing, especially the operations of LeMay's Bomber Command, the unit that perfected massive firebombing on Japan.

McNamara's commanding officer during one assignment said,

> the genius of the operation was the young McNamara, putting all the infinitely complicated pieces together, doing program analysis, operation analysis, digesting the mass of facts which would have intimidated less disciplined minds, less committed minds, making sure that the planes and the crews were readied at roughly the same time. Since all this took place before the real age of computers, he had to work it out himself. He was the intelligence bank of the project, and he held the operations together, kept its timing right, kept it all on schedule. (quoted in Halberstam, 1972, p. 280)

McNamara was only one of many such analysts, and they weren't always welcomed by the Army Air Force. As Michael Sherry notes in his history of the rise of U.S. air power:

> On occasion, the effort to quantify the air war aroused suspicions, particularly among some professional officers . . . [but no] one successfully challenged the general approach whereby numbers measured results, informed rhetoric, and displaced subjective analysis. (1987, pp. 232–233)

Sherry shows that while the military often resisted specific operations research (OR) analysis, and though it certainly challenged the idea of any certainty in bombing, and despite the fact that "military commanders often made critical decisions by seat-of-the-pants methods," the basic assumptions of bomb/destroy/bomb/win, measured in weights and missions was never challenged. This worldview began to colonize the vision of the Air Force officers:

By the language they used, the methods they employed, and the concerns they focused upon, the experts helped change the content of what decision makers in the air force thought about, permitting them to see air war less as a strategic process aiming at victory and more as a technical process in which the assembly and refinement of means became paramount.

They did so in part because the refinement of means and the achievement of destruction were what operations research could most effectively achieve. (p. 234)

Not only was military utility replaced by operational utility, but the military morality of "just war" and "manly combat" was supplanted by an industrial, amoral approach: mass production killing:

The rhetoric and methodology of civilian expertise also defined goals by the distance they interposed between the designers and victims of destruction. The more sophisticated the methods of destruction became, the less language and methods of measurement allowed men to acknowledge the nature of that destruction. A dehumanized rhetoric of technique reduced the enemy to quantifiable abstractions. Statistics of man-hours lost and workers dehoused objectified many of the enemy's experiences and banished almost altogether one category, his death. (pp. 234–235)

By late in the war, there were 400 OR workers in the Army Air Force. They spread their calculations and their technical perspective relentlessly. One key part of this discourse is to refuse to focus on actual enemy deaths, especially civilian deaths. For example, the Committee of Operations Analysis (COA), in its report on the potential of incendiary attacks, made no estimate of casualties. Instead, it preferred to measure the potential damage with other statistics: 180 square miles of urban areas devastated, and 12 million people, or 70 percent of the total population in the 20 cities, rendered homeless (p. 228).

Most analysis was couched in safe language. Destroying residential neighborhoods was called "dehousing." It was

the favorite euphemism for a variety of virtues perceived in an incendiary assault, some spelled out—workers' absenteeism, lower morale, paralyzed systems—some usually left unspoken: the maimed bodies and bewildering toll of the dead. Target analysts recognized that such an assault would inflict scant damage on primary military and industrial establishments. But few questioned the moral or strategic wisdom of the planned campaign. (p. 232)

Even the weather was enlisted in maximizing enemy casualties. Helmut Landsberg, a German-born meteorologist on the 20th Air Force staff, suggested firebomb attacks be timed with cold weather to precipitate influenza and other serious epidemics. Landsberg, or other weather staffers of the 20th Air Force, also called for renewed firebombing in the spring of 1945 because summer was coming and "Favorable fire weather conditions have almost never occurred from June to October at Tokyo" (quoted in Sherry, 1987, pp. 232, 299).

This technological fanaticism, which allowed the Allies to burn whole cities, was not totally different from the fanaticism of the Axis. True, the Allies, especially the United States, preferred to spend machines and bombs instead of men, and preferred destruction from a distance. But the Germans also had missiles, and strategic bombing of their own. Both sides were capable of suicide battles, the United States and Allies more in the beginning of the war, the Axis toward the end of the war. Both sides were remarkably methodical toward their mass destruction ends. Reading about the logistics of the Nazi Holocaust, the working of Stalin's Gulag, and the logistics of the Strategic Bombing campaigns, it is disturbing just how similar they were—especially in their denial of what they were doing. It is true that the Nazi concentration camps killed more people, and on particularly odious grounds, and Stalin's long-running terror probably killed the most. But when one calculates that perhaps as many as 2 million civilians were killed directly and indirectly by Allied strategic bombing, a comparison does become possible.

Strategic bombing, "destruction disguised as technique," is what Sherry calls "sin of a peculiarly modern kind" and which he defines as "technological fanaticism," the product

> of two distinct but related phenomena: one—the will to destroy—ancient and recurrent; the other—the technical means of destruction—modern. Their convergence resulted in the evil of American bombing. But it was sin of a peculiarly modern kind because it seemed so inadvertent, seemed to involve so little choice. Illusions about modern technology had made aerial holocaust seem unthinkable before it occurred and simply imperative once it began. It was the product of a slow accretion of large fears, thoughtless assumptions, and at best discrete decisions. (1987, p. 254)

These "discrete decisions" resulted in the single greatest night of killing in the war, the Tokyo raid of March 9, 1945. Sixteen square miles and more than 84,000 people were burned. The headquarters for that raid was Washington, D.C. Strategic bombing was run from afar. RAF HQ was out in the English countryside, as were the Bomber Commands for the United States. In the Pacific campaign, General Arnold commanded the 20th Air Force from the Pentagon. Wings had bases on various Pacific islands.

The main reasons that control of the burning of Japan stayed in the U.S. capital were to maximize good public relations and to centralize operations research and bureaucratic management. The Army Air Force ran several large public relations campaigns, and it was always careful to release information so that it helped, or at least didn't hinder, the campaign for an independent Air Force.

The ultimate step in strategic bombing was, of course, the atomic bomb. "Working on the bomb instilled a sense of ultimate potency, a triumph of man over nature" (quoted in Sherry, 1987, p. 202). Some scientists and diplomats saw the invention of the bomb as the ultimate weapon that would now finally make modern war impossible. George Kennan, for example, said:

> The atom has simply served to make unavoidably clear what has been true all along since the day of the introduction of the machine gun and the internal combustion engine into the techniques of warfare ... that modern warfare in the grand manner, pursued by all available means and aimed at the total destruction of the enemy's capacity to resist, is ... of such general destructiveness that it ceases to be useful as an instrument for the achievement of any coherent political purpose. (1961, p. 391)

What Kennan didn't realize was that a new type of war would come out of strategic bombing, a type of war symbolized by atomic weapons. Along with this new war, came a new type of militarism that united, for the first time really, science, technology, industry, and war. During World War II, the Nazis and Japanese never mobilized their economy as fully as did the United States and the United Kingdom. After noting this, Sherry goes on to compare Allied industrialism with Nazi and Japanese militarism:

> Economic power was not an alternative to military prowess but another method of expressing it. What happened in the Allied nations was not the decline of militarism or its failure ever to rise in those countries but its transformation. They departed from the path of militarism in the narrow sense that traditional military institutions, elites, and the values associated with them did not dominate. In another, broader sense—the willing enlistment of the broadest array of national energies and elites into the machine of war-making—militarism triumphed in the Allied powers to an exceptional degree, in a variation of it which Alfred Vagts has called "civilian militarism." (1987, p. 193)

Civilian militarism, to quote Vagts, "may be defined as the interference and intervention of civilian leaders in fields left to the professionals by habit and tradition" (1959, p. 193). Vagts noted that often

> Civilians not only had anticipated war more eagerly than the professionals, but played a principal part in making combat, when it came, more absolute, more terrible than was the current military wont or habit. (p. 463)

This is certainly not an uncommon phenomenon. But actually, it seems a little more complex. The military takes on scientific, technological, and management perspectives and the scientists, engineers, and business managers take on military characteristics. Sperry noted how the Army Air Force, especially General Arnold, typified this new synthesis:

> As Arnold's interest in new technologies suggested, what characterized the Anglo-Americans at war was the coalescence, not the divergence, of civilian and military purpose and values. Military elites embraced civilian expertise, civilian elites embraced military purposes. (Sherry, 1987, pp. 193–194)

The first full fruit of this embrace was operations research.

The Spread of Operations Research

> Systems theory is the latest attempt to create a world myth based on the prestige of science. . . . The basic thought forms of systems theory remain classical positivism and behaviorism. As epistemology, it leaves philosophy no further along in resolving the Cartesian dualism; it attempts to resolve this dualism by mechanizing thought and perception, or rather by constructing mechanical models of thought and perception.
>
> —*Lilienfeld (1979, pp. 249–250)*

According to Crowther and Whiddington, authors of *Science at War*, "The development of operational research was one of the chief scientific features of the war" (1948, p. 91). Even more, they claim that "through it, science entered into warfare in a new degree," its major conception being "the reduction of war to a rational process" (p. 119).

It was not a new plan. As far back as Sun Tzu, military thinkers have been trying to rationalize war. Historian Irving B. Holley, Jr., argues that

> history is replete with examples of military commanders—and industrial managers—who have used a form of Operations Research to improve their effectiveness. But not until the era of World War II did OR acquire its elaborate institutional basis and widespread military application, beginning with the Air Ministry unit established in 1937. (1969, p. 90)

It was during World War I, actually, that there was the first systematic application of operations analysis (OA) to military problems. Later, operations analysis was termed operations research, then systems analysis, as it is still called, although now it also blends into cybernetics, game theory, and crisis management. This first application of operation analysis was British mathematician F. W. Lanchester's work, reported in his 1916 book *Aircraft in Warfare* and first applied in the hunting of submarines (Brodie and Brodie, 1973, pp. 271–272).

Lanchester's efforts also marks the beginning of attempts to quantify the deciding elements of battle into "Laws of Combat." Lanchester's equations, and the numerous revisions since then, are at the heart of many war games, strategic debates, and policy decisions. They are based on the assumption that the dynamics of combat can be scientifically analyzed because

> war concerns a comprehensible body of natural phenomena that may be treated scientifically, not merely in a general and qualitative way, but more especially in the specific, theoretical, and quantitative way so effectively pursued in the study of physical phenomena. (Brackney, 1959, p. 30)

Yet, when one looks at the history of the "Laws of Combat" it becomes clear that despite years of research, numerous formulations, and numberless studies the "Laws" are more controversial than ever. They have never been validated empirically or historically and are often flawed by simple mathematical errors as well (Lepinwell, 1987).

The term *operational research* itself can be traced back to 1937 and the British research team working at Bawdsey, the "home" of radar (Crowther and Whiddington, 1948, p. 92). But most of the assumptions go back to Lanchester, as mentioned above. Biologist Solly Zuckerman was an early advocate. He declared that military operations were "an experiment of a crude kind" and that war was a predictable science and not an art (quoted in Allen, 1987, p. 132).

It seems some scientists and soldiers can't help believing that if a problem is given numerical values and framed as an equation or two it is suddenly easier to solve:

> When the scientist examines an operational scheme, he frequently finds that the commonsense view of it is the correct one. But he backs his view by numerical proof. Thus the commander's decision, which may have been correct from the beginning, is transformed from an emotional judgment into an objective fact. In this way, operational research helps to prevent war being run by gusts of emotion, and hunches. (Crowther and Whiddington, 1948, p. 115)

A fine trick to turn emotions into facts. Despite such magic, some of the "experiments" done by the scientists seem less than brilliant. The radar profile of birds was deduced by taking radar soundings of a dead bird hung from a balloon. Trip wires to protect units in the Pacific from Japanese infiltrators were tested "by stringing wires across routes traversed by the large army of women cleaners found in all Ministry of Supply establishments either upright or on their knees with a scrubbing brush," because these "modes of motion were considered typical of Jap movement." Almost all the "experiments" were simple empirical tests; much of the rest of the analysis was paying close attention to details, such as how many rounds of machine gun ammunition were actually used on bomber sorties and counting the number of planes lost due to fire as compared to other damage (pp. 106–113).

The limitations of this kind of thinking are clear from looking at the work of Dr. Zuckerman and his colleagues J. D. Bernal and E. R. Garwood in predicting damage and casualties from strategic bombing raids. They bombed some goats, observed some German bombing results, and ran some statistical analysis until they could predict, with what enthusiasts describe as "a considerable degree of accuracy," the casualties and other effects per ton of bombs. This work became a major justification for Britain's own indiscriminate and supremely ineffectual night bombing of Germany. As it's described in the semi-official history, "This feat gave the conception of a scientific bombing attack on Germany a new degree of reality and accuracy" (pp. 98–99).

Reality, yes. It helped make the bombing a reality. But accuracy? Scientific? Nothing could be further from the truth. Yet, "these investigations in 1940–41 became a guide to future bombing policy" (p. 99).

Operations research did enjoy some success in improving antiaircraft aiming, in coming up with antisubmarine strategies, and in solving production and logistical problems. But when applied strategically, such as in the bombing of Germany and Japan, it was a failure. But it was successful enough culturally to lead to its institutionalization at the end of the war:

> After . . . World War II . . . the services . . . committed to long-term operations research analysis. The Air Force organized the Rand Corporation . . . the Army established the Operations Research Office . . . the Navy . . . continued its war-time operations research team. . . . At the level of the Department of Defense, a Weapons System Evaluation Group was established, along with an independent counterpart, the Institute for Defense Analysis. (Janowitz, 1971, p. 30)

This was just the institutionalization of a more fundamental shift in the postmodern military. The historical attitude of the military toward new technology, that of conservative resistance, had by the end of World War II swung 180 degrees so that now, in Janowitz's words, "the arms race in nuclear and guided weapons has converted the armed forces into centers of continuous support and concern for innovation" (p. 27).

This change had but slight beginnings. In December 1942, General Arnold, the Deputy Chief of Staff of the Army and head of the Army Air

Force, authorized a Committee of Operations Analysts to work with Air Staff. This committee was dominated by lawyers and businessmen: Elihu Root, Jr., Thomas Lamont of J. P. Morgan and Company, John Marshall Harlan, later of the Supreme Court, Fowler Hamilton, and the economist Edward Mason (Holley, 1969, pp. 189, 194). During the war, as mentioned above, there were over 400 such analysts.

After World War II, the headquarters OR staff was limited to 10 under LeRoy Brothers, and General LeMay's Strategic Air Command (SAC) had 15. Counting the other Air Force OR units as well, Brothers hired 6 physicists, 5 mathematicians, 23 statisticians, 7 engineers, and 1 education-alist (Holley, 1969, pp. 57–58, 93). But many more analysts were employed in outside think tanks, such as RAND. There was some struggle at first over the scientific status of OR, but the lawyers and even the social scientists were soon purged from the discipline:

> One important element shaping the destinies of the Air Force OA organization was the Operations Research Society of America, commonly called ORSA. The founding of ORSA in 1952 clearly reflected the increasing appreciation for the work of analysts in both military and industrial circles. From its infancy, the Society favored a "hard science" approach. (p. 95)

The growth of cybernetics, which began to jell as a discipline during the war,[7] backed by computers, lent further support to the "hard science" approach. Ellis Johnson of the Army-sponsored Operations Research Office, affiliated with the Johns Hopkins University, urged that social scientists be included in OR, but he was in a minority. "After virtually ignoring the tool during World War II" the Army was spending 10 million a year on OR by the mid-1950s (pp. 94–95). The moderately successful operations analysis of World War II, with teams of lawyers, engineers, and scientists, became much more scientistic and much less successful after the war. In a way, it went somewhat "insane," as the history of strategic doctrine will show in the next chapter.

Within the military domain, the scientists set higher goals. Instead of judging weapons, the scientists decided they should make the tactics and the strategies, to think through the whole system and make a systems analysis (SA):

> Where the analysts of the war years had worked to make the performance of *existing* weapons optimal, the task now was to make *future* systems more effective. So the emphasis shifted to studies of alternative strategies, to the determination of requirements posed by these strategies, and to the promulgation of the detailed performance specifications demanded of each weapons system by the strategy selected. (p. 100; emphasis in original)

Unsurprisingly, World War II–style OR basically missed out on Vietnam, a war framed by SA. For a brief period in 1968 some OR teams got involved in analyzing appropriate B-52 sortie rates and the B-52's use as a tactical weapon. Since they couldn't analyze bombing effectiveness, they got into target selection (pp. 106–107). But how can you tell what is a good target to aim at if you don't know which targets you can hit? OR was just too practical for Vietnam, a war that was not interested in how things were but how they were supposed to be.

Along with the systematic calculations of OA, OR, and SA, a whole new type of machine evolved out of World War II to help the human calculators—mechanical computers.

The Spread of Computing Machines

The very first attempts to build working computers, the calculating engines of Charles Babbage, were paid for with the very first military research and development grant—this one from the Royal Navy. While Babbage's machines never worked, a modified Babbage Difference Engine, designed by Georg and Edvard Scheutz, did produce some of the astronomical tables Babbage was hired to make. Charles Xavier Thomas de Colmar "invented, perfected, and manufactured and sold the first commercially successful digital calculating machine" around the same time as Babbage worked. It was called the Arithmometer, and his customers included the French and British armed forces (I. B. Cohen, 1988, p. 125). Since these beginnings the connection between computing machines and military needs has been incredibly intimate and symbiotic.

By the end of the nineteenth century simple mechanical calculators were common. On land they were used by the artillery; on sea they directed naval guns. The workhorse of calculators for the United States during World War II was probably the analog Differential Analyzer, invented by Vannevar Bush. Two of them, built by the Moore School of the University of Pennsylvania, were installed at the Army Ordnance Department Ballistic Research Laboratory at the Aberdeen Proving Ground in 1934 (p. 135). But it wasn't until World War II that the first true computers were developed, and it was at the military's behest.

George Sitbitz, who built a protocomputer in 1939, describes the effect of the war on the development of real computers:

> While the demand for rapid and inexpensive computation had been increasing for years before the war, the increase was, as we look back at it now, slow and insignificant in comparison to that which the war engendered. . . . New military devices, unsolved problems of tactics, unfamiliar

subjects—they all demanded mathematical treatment if possible. (quoted by I. B. Cohen, 1988, p. 135)

What some people have called the first actual computer[8] was built just as World War II was starting. It is almost unique among the early machines in that it was not paid for by the military. However, its designer, John Atanasoff did end up working for the U.S. Navy, and the working machines that came after his prototype were all built for the military (Loevinger, 1989). The point is not that computers would never have been developed without the military's interest; after all they are an old dream. But they certainly would not have been invented so quickly, built so fast, nor designed in the ways they were without the sponsorship of the military.

Once the war was going in earnest the military put a tremendous effort into building more and better computers. William Rodgers notes in his history of IBM, *Think*, that the theory and technology to make the early computers, the ENIAC (Electronic Numerical Integrator and Computer) and the Mark I, was in place "twelve to fifteen years earlier" but there was not "the money or . . . incentive to do it" until the Army stepped in (1969, p. 189).

Actually the U.S. Navy and Army Air Force as well as the British military all played important roles in the birth of the computer. Probably the first real programs, and many other key aspects of modern computing, were developed by the British cryptanalysts, led by Alan Turing, with their Enigma machine. They played an important part in the Allied victory, but so did other computers—men and their machines.

Almost all the other founding fathers of cybernetics and computer science besides Turing were also doing war work. Norbert Wiener, who coined the term *cybernetics*, spent the war working at the MIT Radiation Labs. Vannevar Bush, an early experimenter on analog computers, played a major role managing military science as the head of the National Defense Research Committee. Claude Shannon did work on secrecy systems.

Wallace Eckert's work for the Naval Observatory with IBM machines produced the first sequence-controlled calculator, which was used to generate navigational tables crucial in antisubmarine warfare. Prof. Howard Aiken built one of the world's first computers, the Mark I, for the U.S. Navy at Harvard. His machine, and another at Bell Telephone Laboratories, were used to calculate trajectories of heavy artillery and antiaircraft guns. The Mark I also calculated the design of lenses for reconnaissance photography and did applied mathematics dealing with magnetic mines. IBM built a number of Pluggable Sequence Relay Calculators for producing ballistic tables (Ceruzzi, 1989a, p. 13; I. B. Cohen, 1988, p. 131). There was also the computing work on the atomic bomb and ENIAC, the first electronic digital computer, which helped design the hydrogen bomb (Rogers, 1969, pp. 155–186). Although, as Paul Ceruzzi points out (p. 13) computers were not crucial for making the first atomic bombs, the Mark I did calculations for John von Neumann on the problem of implosion (I. B. Cohen, pp. 131, 134). Computers did become central to the design and manufacture of all future nuclear weapons. The first problem ENIAC calculated was a physics equation for the atomic bomb laboratories at Los Alamos (Edwards, 1987, p. 51).

John von Neumann, considered with Turing, Wiener, and Shannon central to the creation of computing, spent most of the war at Los Alamos helping with atomic bomb calculations. It was there he also helped create game theory. But he did find time to consult on the Army project to build ENIAC at the University of Pennsylvania in Philadelphia (Heims, 1980, p. 181).

Radar, one of the truly decisive technologies of World War II, contributed many of the scientists and engineers who made computers possible, and at least one company—Raytheon. As Werner Bucholz remarked, "many computer projects started around a nucleus of wartime radar experts" (quoted in Ceruzzi, 1989b, p. 259).

Radar was part of the work on gun control that led to the founding of the Servomechanisms Laboratory at MIT to develop remote control systems, gun drives, radar, and similar machines. From that its work spread to analog and then digital computers. By the end of the war it was working on Whirlwind, a programmable flight trainer, for the Navy. This was a key beginning point for the postwar attempts by the military to apply computers and robotics to both war and industrial peace (Noble, 1986b, pp. 106–143).

Electromechanical analog computers predicted the shell trajectories for battleships' 16-inch guns. They not only took into account the direction and elevation of the guns but also the rotation of the earth and the temperature of the propellant. Antiaircraft guns used "predictors" to aim, and bombers used cathode ray tubes to record when complexes of radar stations (called masters and slaves) put them over their targets, including the fire bombing of Hamburg (Crowther and Whiddington, 1948, pp. 51–55). The famous Norden bombsight includes a built-in calculator to gauge the influence of the plane's speed, altitude, and drift.

Gun control led to what was probably the first attempt to develop an autonomous weapon, the automatic gun-laying turrets (AGLT), for Bomber Command. Freeman Dyson, who worked on this project, explains how the research engineers came up with a "magnificent solution" involving linking a "high-performance search-and-track radar and a gyroscopic gun sight . . . to a servosystem which aimed the turret and guns." While the attacking fighter was still out of sight, the gunner could pull the trigger and destroy it. After a crash program the AGLT was produced and readied for operations, but it couldn't be installed until every Allied aircraft had an automatic IFF (interrogation friend or foe) unit so the AGLT wouldn't target friendly fliers and commit what the military calls fratricide. IFF accuracy had to exceed 99 percent if the weapon was to kill more enemies than friends. It never came close. After much effort at the height of the war, the AGLT had to be abandoned. To this day the problem of fratricide and IFF has meant that the vast majority of air combats, including the most recent, have involved visual sightings, often at subsonic speeds, with machine guns or cannon being as important to combat success as missiles (Dyson, 1984, pp. 57–60).

The remote control of weapons, however, was advanced during the war. It started with a 1942 British project to use radar to guide bombers, termed "Bombing without Knowledge of Path, Place or Time" (Crowther and Whiddington, 1948, pp. 58–59). Germans had radio-guided bombs such as the HS-293 and the Fritz X, while the United States recycled old B-17s, filled them with 10 tons of explosives, named them Weary Willie's and sent them, radio-controlled by a manned B-24, with little effect against German submarine pens in the North Sea (Editors of Time–Life, 1988, pp. 62–63). These planes were referred to as "robots," "drones," and "missiles" by the Army Air Force, which was also working on an "orphan" version to be controlled from the ground with a range of 1,500 miles.[9] There was also a plan to use crewless, explosive-laden boats controlled by aircraft beyond the range of vision. These Javaman boats, as the joint AAF–OSS project named them, would carry television cameras which the aircraft would monitor in order to guide them. They were approved for action against Japanese interisland railway tunnels, but the war ended before they could be tried.[10]

The Germans had more success with remote weapons. The HS-293 guided bomb and the SD-1400 gliding bomb were used to sink several ships, including the Italian battleship *Roma* while it was on its way to accept the Allied surrender terms. The Messerschmitt ME-163 rocket fighter, the FZG-76 (V1) pilotless flying bomb, and the A-4 (V-2) long range rocket all worked pretty well. Wernher von Braun also developed the beam-guided *Wasserfall* missile with a speed of 1,700 miles per hour and range of 16.5 miles, and Dr. Max Kramer invented the *Ruhrstahl* X-4, a 520-miles-per-hour air-to-air missile with a 3.4-mile range guided by two fine wires played out from the missile wings, and the X-7, a similar antitank missile (Macksey, 1986, pp. 158, 163, 168–169). When the war ended the Nazi scientists were working on an intercontinental rocket, the A-10, with a range of 3,500 miles. It would have been capable of hitting New York City and was scheduled for use in early 1946 (Sherry, 1987, p. 120).

It is a little known detail of technoscience history that two Germans, Konrad Zuse and Helmut Shreyer, almost invented the first computer. Zuse's first prototype was financed by Kurt Pannke, a manufacturer, but his later machines, and those of Shreyer, were paid for by the military directly or indirectly through the Aerodynamics Research Institute. Because of the war, which prevented funding of some projects and also resulted in the bombing of a number of prototypes, very few working machines were built. Still, Shreyer built the first electronic circuits and Zuse made several advanced calculators. The Z3 may have been the first programmable machine but it was never put into real use because of memory limitations. The S1 took over 100 wing measurements for cheap unmanned flying bombs and calculated their future aerodynamic performance. When it was installed it replaced over 30 women "computers." The Z4 wasn't completed until the very end of the war. It was put into operation in Switzerland in 1950 and for several years was the only computer in Europe (Ceruzzi, 1983, pp. 21, 28, 38–39). The Germans had other mechanical computers as well, which they obviously felt were of supreme military importance, as the following strange incident at the end of the war in Europe revealed.

When the Third Reich fell U-234, a German submarine, surrendered itself at Portsmouth, N.H. Her mission, never fulfilled, was to deliver key technologies and information to Japan, Germany's main ally. Although the two Japanese officers on board killed themselves after writing a moving message of farewell to their German colleagues, the complete cargo and documents of U-234 were captured. Obviously, space on a submarine going half way around the globe was limited, so it is indicative of the growing importance of computers that a number of such machines, for both cryptography and fire-control calculations, were carefully stored on board along with various aircraft and missile blueprints, electronic circuits, and bars of platinum.[11]

Despite Gen. Dwight D. Eisenhower's order that "There'll be no software in this man's Army!"[12] the postwar period was marked by an explosion in computer development, much of it paid for by the military. The military also facilitated a series of key conferences, such as the 1947 and 1949 meetings at Harvard, cosponsored by the Navy Bureau of Ordnance, and the 1947 and 1948 meetings at the Moore School, cosponsored by the Navy Office of Research and Inventions and the Army Ordnance Department. In the United Kingdom, the Ministry of Supply sponsored several key conferences as well, and the Post Office built a computer, MOSAIC (Ministry of Supply Automatic Integrator and Computer), for them as well (I. B. Cohen, 1988, pp. 123–124). Between them, the U.S. and British military paid for a huge proportion of early computer development.

The U.S. Navy's Special Devices Division, for example, which was started during the war to "exploit advances in electronics and other technologies," initiated a series of computer projects, Whirlwind, Cyclone, Typhoon, and Hurricane, each of which "simulated an aspect of air or space flight" (Ceruzzi, 1989a, p. 249). In the ten years from 1946 to 1955, the Office of Naval Research (ONR) cosponsored Whirlwind I at MIT, IAS (the Institute for Advanced Study), NAREC at George Washington University, and other hardware research at the National Bureau of Standards. They also paid for mathematical studies of fluid dynamics, plasticity theory, nonlinear control mechanisms, linear programming, game theory, decision theory, and inventory management. Just as important as this research, was ONR's regularly published "Surveys of Digital Computers" and the *Digital Computer*

Newsletter. These publications were the first to network the computer industry/community together (I. B. Cohen, 1988, pp. 141–143).

In 1946 the airplane manufacturer Northrop, working for the U.S. Air Force, asked J. Presper Eckert and John Mauchly, coinventors of ENIAC (Electronic Numerical Integrator and Computer) to build the first compact computer, BINAC (Binary Integrator and Computer) to control the Snark, a jet-propelled, swept-wing, pilotless bomber. While they failed, the contract kept their company, Electronic Control Corporation alive and BINAC was the first operational computer in the United States to employ the stored-program principle. Northrop then hired Hewlett-Packard to build MADDIDA (Magnetic Drum Digital Differential Analyzer), which, although more reliable than BINAC, was also too large for the Snark, which ended up with an analog machine instead.

The Air Force also launched Project SCOOP (Scientific Computation of Optimum Problems), which funded George Dantzig while he invented the simplex method of linear programing to solve difficult logistical calculations. To run his programs the Air Force had the National Bureau of Standards build SEAC (Standards Eastern Automatic Computer), completed in 1950. But it really wasn't powerful enough for linear programing, so in 1952 the Air Force brought the second UNIVAC (Universal Automatic Computer) made by the Electronic Control Corp. (Ceruzzi, 1989a, pp. 24–25, 41).

The next year, 1953, the Air Force think tank, RAND, built its own computer, the JOHNNIAC, named after John von Neumann.[13] That same year IBM began shipping its 701 computers, originally named the "Defense Calculator" (I. B. Cohen, 1988, p. 139). After installing the first 701 in their own world headquarters, IBM sold 701s to Los Alamos, Lockheed, the NSA, Douglas Aircraft, General Electric, Convair, the U.S. Navy, United Aircraft, North American Aviation, RAND, Boeing, another to Los Alamos, another to Douglas, another to the Navy, Livermore Labs, then General Motors, another to Lockheed, and finally, the nineteenth, for a joint military–Weather Bureau project (pp. 44, 46).

In the 1960s, massive computing projects continued. Most notably the SAGE air defense radar system of 52 linked Q-7 computers. It became operational in 1963. The system cost $8 billion dollars, which was a lot of money in the early 1960s (Ceruzzi, 1989a, p. 71). Aircraft, air defense, missiles, and nuclear weapons development continued to drive computer science as they had from the beginning. The Minuteman ICBM project, for example, "nurtured integrated-circuit production through its infancy in the early 1960s." Even in 1965 Minuteman missiles still represented 20 percent of the integrated-circuit market. Ceruzzi concludes that the Minuteman II project played "a central role in bringing the chip to its present position in society" (1989a, p. 94).

The very first customers for Cray-1 supercomputers were the atomic energy organizations of the United States and the United Kingdom, which used them to design nuclear weapons. While veiled in secrecy, it is clear that the massive nuclear weapons industry has always been a major consumer of computers and a major producer of various innovations, especially in software.

Although it is equally shrouded in secrecy, there is also evidence that during this period the secret agencies, especially the National Security Agency, played a major role in the development of the computer. By 1990 the NSA had acres of computers at Fort Meade, Md. That was the result of years of effort, such as Project Lightning, implemented in 1957 by then NSA Director Bernard Canine, which aimed at increasing the speed of computers more than 1,000-fold in just five years. It was one of the few nonsecret NSA initiatives. After spending $25 million the NSA could claim a number of hardware improvements and partial credit for keeping several young computer firms alive. The NSA still stands high in the computer pecking order; it still usually buys one of the first production models of any new Cray or other supercomputer (Editors of Time–Life, 1988, pp. 46–47).

* * *

Strategic bombing, systems analysis, and computers were the great military victors of World War II. It wasn't that they won the war, but they won the peace. They are reoccurring themes. Whether successful or not, the AI paradigm has continued to win converts within the military and outside it, in a reinforcing dynamic that shows no signs of slowing down. Indeed, these trends from the world wars have blossomed in the last 50 years into the system of postmodern war we now live with. They have become metarules in war's discourse system.

Seeing war as a discourse system, as well as a military reality, might help us understand why thousands of soldiers were sent to certain death against machine guns in World War I and why millions of civilians were bombed to death in World War II *for no good reason whatsoever.* In World War I the dominant discourse metarules insisted that attacking machine guns was the road to victory. Hundreds of thousands of deaths and several new technologies (tanks, gas) were produced before the faith in force of will was broken and faith in technology took its place. That faith in technology assured military technophiles that bombing cities would lead to victory, despite all the evidence to the contrary. The discourse metarules that justified strategic bombing are behind the current computerization of war that is the core of postmodern war. Through this discourse these metarules have determined the planning for postmodern wars, from the horrific imagining of World War III to the real horror of Vietnam.

Chapter Eight

Postmodern Wars Imaginary and Real: World War III and Vietnam

Computing the Unthinkable: Planning World War III

The nuclear weapon and the AI mentality represent two different ways of looking at the lessons of World War II. Was war now too horrible to wage? Or was it now manageable, thanks to science and technology? At first it seemed the nuclear weapon metaphor would dominate. Many people, including military officers, declared an end to war.[1] Abolition was proposed by a few policy makers, international control advocated by even more. But soon the AI mentality became the dominant frame of reference. Resistance to the idea that war was still an integral part of politics has continued to this day but has been marginalized from the central discourse. James Skelly explains the regime of truth that has developed:

> The techniques and procedures accorded value in the acquisition of truth with regard to deterrence are in general those of political science, game theory and system analysis. The status of those charged with saying what counts as true generally depends upon institutional position. (1986, p. 10)

At first deterrence was simple. Bernard Brodie pointed out winning a total war against a nuclear opponent was impossible. There could be no winners. The bomb wasn't a military weapon—it was political (1946). Then, from the new centers for operations research (OR), systems analysis (SA), game theory, and gaming itself came new strategic doctrines that needed new nuclear weapon systems with new delivery technologies. These theories were applied with disastrous results during the U.S. intervention in Vietnam (Vickers, 1984; Wilson, 1968). Meanwhile, atomic bombs were replaced with hydrogen bombs. Bombs were supplanted with missiles. Land missiles were augmented with submarine missiles. "Mutually assured destruction" (MAD) was superseded by counterforce doctrines of controlled response. Strategic nuclear weapons were joined by tactical ones. All of these expansions were justified by elite research institutes.

Jonathan Schell comments on an important reason that strategic theory was relegated to these "think tanks":

> This term, evoking a hermetic world of thought, exactly reflects the intellectual circumstances of those thinkers whose job it is to deduce from pure theory, without the lessons of experience, what might happen if nuclear hostilities broke out. (1982, p. 140)

U.S. strategic doctrine sprang from the think tanks founded after World War II, especially RAND Corp. In *Wizards of Armageddon*, a history of strategic doctrine, Fred Kaplan (1983) traces the military roots of most of the key theorists:

- RAND founders Arthur Raymond and Larry Henderson were with the B-29 Special Bombardment Project and the Office of Scientific Research and Development.
- Albert Wohlstetter worked for the Planning Committee of the War Production Board doing OR and scientific management. At RAND he did SA and game theory.
- Ed Paxson invented SA, by which he meant asking which system did the job best instead of asking how can this system (operation) be made to function most efficiently. This meant becoming a military planner instead of being just an efficiency expert. The terms are now almost interchangeable except among adepts. Kaplan describes Paxson's fascination with planning World War III:

> His dream was to quantify every single factor of a strategic bombing campaign—the cost, weight and payload of each bomber, its distance from the target, how it should fly in formation with other bombers and their fighter escorts, their exact routing patterns, the refueling procedures, the rate of attrition, the probability that something might go wrong in each step along the way, the weight and accuracy of the bomb, the vulnerability of the target, the bomb's "kill probability," the routing of the planes back to their bases, the fuel consumed, and all extraneous phenomena such as the weather—and put them all into a single mathematical equation. (1983, p. 87)

These men were joined at RAND by Herman Kahn, who became an SA expert and then applied "his peculiar brand of logic, mathematical calculation, [and] rationality" to the fighting of nuclear war (Kaplan, pp. 221–226).

Detailed histories of U.S. strategic doctrine show that many factors (ranging from political to technical) have helped shape it. Among the most significant was, and is, the use of scientific justifications such as SA and game theory. The careful hunt for "worst case" scenarios, combined with the inevitable bias toward measuring enemy capabilities (usually much exaggerated) instead of opponents' intentions, meant the extraordinary proliferation of nuclear weapons, nuclear platforms, and nuclear missions.

In his book *Think Tanks*, Paul Dickson credits RAND with numerous innovations in these areas in its attempts to answer what RAND itself called "the unsolved problem of modern total war." He lists scenario writing, heuristic programing, computer modeling, linear programming, dynamic programming, problems scheduling, nonlinear programming, the Monte Carlo method, the Planning, Programing and Budgeting System (introduced in the DoD in 1960 and the rest of the federal government in 1965), and SA itself (1971, pp. 61–64). Most of these techniques were borrowed directly or metaphorically from computer science. What makes them particularly interesting—and postmodern—is that they are all ways of simulating (gaming, predicting, calculating) possible futures so as to justify constructing one in particular. Their effectiveness can only be judged by their results. They have supplied the rationale for the growth of the U.S. nuclear arsenal from 40 weapons to over 25,000.

Defense Secretary Robert S. McNamara admitted that U.S. nuclear policy has been "war-fighting with nuclear weapons" from the beginning (quoted in Herken, 1985, p. 306). The original Strategic Air Command (SAC) plan of massive attack on the Soviet Union was soon attacked itself by the RAND consultants. Using systems analysis, Albert Wohlstetter claimed he had proven that the Soviets could "Pearl Harbor" the U.S. nuclear bombers. This threat and a parade of others usually labeled "gaps," as in the bomber gap and the missile gap, have all been shown to have been nonexistent—after the election was won or the new weapons were built. But before being proven fallacious, they helped to create and elaborate the nuclear arsenal of the United States.

Other think tank contributions to fighting nuclear war include RAND's development of the "distributed network concept" to make a communication system capable of continued operation during a nuclear attack and the development by a RAND spin-off, System Development Corp. (SDC), of the software for various air defense systems, including work on SAGE (Semi-Automatic Ground Environment) and BUIC (Back-Up Interceptor Control). One of the first of the 1,000 military jobs performed by SDC involved more than 500 programmers writing over a quarter of a million computer commands for the nation's air defense system, the biggest program of its time. SDC also designed the Navy's Space Surveillance System, and it has produced numerous nuclear war games for the Pentagon. The Pentagon's in-house think tank, the Institute for Defense Analysis (IDA), has done many studies on fighting nuclear war, including a series on reestablishing the U.S. government and economy after nuclear war. The Army's think tank, HUMRRO (the Human Resources Research Office), has contributed research on how to help soldiers fight after they've been attacked by nuclear weapons. Another key Army think tank, Research Analysis Corp., has developed scores of computer war games and also helped establish the principles for the tactical use of nuclear weapons (Dickson, 1971, pp. 71–151 passim).

Whatever the catch phase—counter-force, no-cities, massive retaliation, assured destruction, finite deterrence—it always meant more nuclear weapons. The justification was always framed by SA, game theory, bargaining theory, and reams of numbers on computer printouts. The number crunching wasn't just in the planning and modeling aspects either. The SA and mathematical manipulation of the collected intelligence on the Soviets was crucial for opening the missile gaps and windows of vulnerability necessary to bring new weapons into reality.

Jim Falk explains something of the underlying ideological structure of this discourse system:

> Whether it is in relation to strategic theorizing, strategic studies, or more generally strategic discourse, the analysis here suggests that the dominant discursive practices have been selected around an ideology which emphasizes the desirability and attainability of successfully modeling, predicting and controlling human affairs. (1987, p. 19)

This desire to model, predict, and control human affairs, especially warfare, is never satisfied, especially in the horrific realms of nuclear war. Among nuclear theorists and war planners this leads to some interesting psychological reactions. Steven Kull, a practicing psychotherapist, interviewed scores of policy makers in an attempt to discover something of the psychological processes involved.

In his book *Minds at War*, Kull (1988) quotes many top officials to the effect that nuclear weapons are not to be used, but at the same time the same men keep a commitment to fight nuclear war. According to Kull, these two strands of nuclear thinking—that nuclear weapons are for deterrence or they're for war fighting—coexist in the minds of Soviet/Russian policy makers and of most American policy makers.

On the one hand, such officials could be very clear about the impossibility of fighting a nuclear war. Said one State Department officer: nuclear weapons are "to make yourself feel that you're doing your part." They aren't just to impress the enemy but "you build to impress yourself. . . . We're building this so . . . we'll feel better." A high-level Pentagon official said that his work on nuclear strategy was "quite schizophrenic," based mainly on bravado and public relations, not "serious . . . military possibilities." Another State Department official admitted that Reagan policy for winning a nuclear war was not a real policy but was only really established "for perceptual reasons." It was meant to impress allies as much as enemies (Kull, 1988, pp. 232–233).

On the other hand, most of the officials share the assumption that there will always be war. A Pentagon official: "[Nations] want to fight . . . [because of] these genetic factors and Oedipal factors and everything else like that." A presidential adviser: "My whole theory of history is based on patterns. . . . The most important pattern about war can be stated in two words: it's recurrence." Both stress that despite war being irrational, it will still happen. "Wars are just going to go on," claimed one. "For people to act rationally goes against all history," summarized the other. A State Department official put it more starkly:

> Tensions begin to build up with you and [leaders] start making decisions to release the tension . . . [so] at a certain point everyone just decides, "Fuck, let's go to war!" It's just easier . . . I mean, you're not worried about, well, should we, shouldn't we do it. It's not a rational decision. (quoted in Kull, 1988, p. 239)

A main driving force of this logic is technology itself. As Kull notes, "several respondents said, in nearly reverential tones, 'You can't stop technology!' "[2]

Kull explores in detail how every rationale was deployed to justify the nuclear arms race. A number of interview subjects admitted they couldn't justify their desire for more weapons when overkill was already surpassed ten times over. It was just a "visceral" need. Others, such as Colin Gray, justified more weapons because an arms race was less violent than actual war. Kull concluded that his subjects had a conception of humans as competitive and warlike and they (the subjects and all humans) could not go against what they thought was their very "nature of being." He adds:

> At the deepest level it seemed that the most fundamental motive was almost mystical in nature: the desire to align oneself and fulfill not only one's own deepest nature but the deepest nature of being itself. (p. 247)

Kull argues that these men managed to maintain emotional equilibrium by either suppressing thoughts and emotions or rationalizing them. A major factor in nuclear discourse is prenuclear "conventional principles," often evoking as emotion-laden symbols such historical events as Pearl Harbor, the appeasement of Hitler, or the Soviet's Great Patriotic War. All three of these historical tropes is an argument for a level of military preparedness that is probably inappropriate for a nuclear arms race. This emotional logic is very compelling. Almost all of Kull's respondents admitted under his questioning that the policies of nuclear warfare they were proposing were questionable or even invalid. But they would return to more traditional military calculations and rhetorical tropes from modern war as the interview wore on. It seemed traditional defense policies were just too psychologically gratifying to resist. Kull traces this back to the very roots of war as "ritual . . . that involved an enactment of fundamental mythical and religious themes" (pp. 302–307). He goes on to show (pp. 308–312) the religious bases for nuclear policy, with a quote from Richard Nixon on the need for the "drama" of good and evil, and with a survey that counted 35 percent of all members of Congress agreeing that "God has chosen America to be a light to the world." Kull found that the assumption that nations must compete and therefore, perhaps tragically, there will be war were ontological beliefs, beyond question for most of his subjects. His frightening conclusion was that

> [such] ontological beliefs, to play a critical role in shaping behavior, do not even have to be fully conscious. As fundamental assumptions they form the bedrock of culture from which arise values and ultimately concepts of the national interest. The desire to behave in ways that are consonant with these deeply held structures is quite powerful and, as has been demonstrated again and again over the centuries, can be more fundamental than the desire to survive. (p. 310)

The Reagan administration took such feelings to their limits. Theorists such as Colin Gray openly talked about winning a war through decapitation (nuking the other side's leaders, presumably by a preemptive first strike). President Ronald Reagan dreamed about ending the threat of war, at least for the United States, by hiding behind the SDI shield. Meanwhile, in deepest secret, his administration made plans not only for fighting and winning World War III but World War IV as well! Every possible justification for weapons seemed to have a place (Scheer, 1982). Under President George Bush the fascination was with midintensity and low-intensity conflict, but waging nuclear war was never renounced. All avenues that might lead to war were kept open, but the low road is the easiest traveled.

A Genealogy of Little Wars

In the beginning all wars were little, very little. They were fought between dozens or perhaps hundreds of men. Only in the last 5,000 years have wars been a matter of thousands of men in armies confronting each other. Since then, most "little wars" have been when ancient or modern armies fought rebellions, confronted nomads or met ritual warriors in unequal combat. Irregular, "guerrilla" forms of resistance did often hold off ancient armies, when the cost seemed too high or the gain too small for the empire builders. But seldom could primitive warfare defeat modern armies.

In the modern era resistance to Western expansion was tenacious, often lasting hundreds of years. However, with very few exceptions (Haiti, Ethiopia, Afghanistan, Japan) the West always triumphed eventually. These wars were marked by the use of surrogate forces, the effective deployment of new technologies, and incredible moral license. In many cases genocide was the strategy for victory. These colonial conflicts were called "imperfect wars."

Capt. Bernardo De Vargas Machuca, who campaigned in Chile, wrote the first manual on combatting guerrilla warfare in 1599. He advocated using commando groups on extensive search-and-destroy missions of up to two years to exterminate the Indians. Native tactics, such as living on the land, ambush, and surprise attacks, were to be employed by the Spanish as part of their strategy of extermination. Such wars ended with an extended manhunt as the last "wild" Indians were tracked down with dogs and killed in cold blood (Parker, 1988, p. 120). Bloodhounds were still being used to track Indians as part of a successful strategy in the Florida wetlands and California hills in the nineteenth century.

But at the end of World War II a sea change took place. Colonialism was suddenly collapsing under political and military pressures that ranged across the spectrum from *satyagraha* (the "truth force" of Mohandas K. Gandhi's independence movement) through voting to violence. Part of the great rollback of colonialism was because of the obvious superiority, for the West, of neocolonialism, as exemplified by the United States's domination of Latin America and the Philippines. Other nations thrust off European sovereignty only by ceding some measure of control to regional powers or to the so-called Second World. The Communist bloc certainly did not lead countries to liberation, but it helped create a space for room to maneuver, if not real independence (Chaliand, 1978). This is seen most clearly in the case of war. Consider the spectrum of allies the Vietnamese mobilized: Sweden, China, the USSR, and large parts of the international peace movement.

While modern war was developing toward total war, it was often challenged by irregular war. When Western armies met each other, they fought modern wars, but when they went to war against other people it was a more limited struggle, from the point of view of the Europeans. They called it insurrection, revolution, guerrilla war, tribal revolt, rebellion, uprising, police action, little war, imperfect war, colonial war, and limited war. These wars were only limited on one side. For the nonindustrialized societies they were total wars sure enough, often leading to the destruction of whole cultures and the genocide, or near genocide, of entire peoples. Postmodern war has imposed the framework of "minor" wars onto all conflicts, even those between the great powers. This came as something of a surprise to them, witness the stalemate called the Korean War.

The Korean experience . . . impelled the development of the concept of
"limited war," the notion that the major countries could wage a war that
could be restricted in geography, weaponry, and goals. . . . If war can be
limited, it may become "possible" again: if the combatants tacitly or
explicitly agree to stay within tolerable constraints, they can test national
will and military prowess while keeping damage within bearable bounds.
(Mueller, 1989, pp. 129–130)

Still, a limited war can be very bloody. Over 2 million died in Korea. But
considering the butcher's bill of a full war between China and the United
States it is clear that it was limited for them, if not for the Koreans. These
nontotal wars can range from skirmishes to a serious conflict, now labeled
midintensity. In U.S. planning, such limited wars were seen as police actions
or at the most half wars, worth half of a total war with another superpower.

But even before fighting raged up and down through the Korean
peninsula the first postmodern war began in Vietnam. What makes this war
so important is that it reversed the hundreds of years of European victories,
and not just in one battlefield but in the minds of millions of people around
the world. True, indigenous people had won many battles in the past, but
they had lost almost all the wars. And, true, Japan had defeated Russia in
1905, but Japan was an industrialized power and it had never been colonized.
In the long Vietnam struggle, a small agricultural country defeated first
France, its colonizer, and then the most powerful empire in history, by
framing war in basically political terms.

This was not a totally new strategy. In part, it was George Washington's
approach during the American Revolution, as the Vietnamese well knew. It
also was an approach that drew strongly on the hundreds of years of Vietnamese resistance to colonialism and on the more recent experiences of the
Chinese Red Army. But, granting all that, it was still incredible. Politics
became war by other means for the Vietnamese. Not that military skill and
courage were not needed; they were crucial to keep the struggle going until
the eventual political exhaustion of the invaders. The theory of "people's
war," as the Chinese and Vietnamese call it, was so sophisticated that it even
laid out the transition between the lowest levels of military resistance on
through to eventual victory through conventional confrontations.

Some have argued that people's war is necessarily ideological, but the
last 50 years show it isn't, at least in a left–right sense. Not only have the
Algerians used it successfully, but so have the Afghans. And, with bittersweet
irony, more recently the same type of war has been turned against some of
its best earlier practitioners in Southern Africa (Angola and Mozambique)
and Southeast Asia. For example, in Cambodia the Vietnamese themselves
could not crush the Cambodian resistance fighters, some of whom were
supported militarily by China (the Khmer Rouge), others covertly by the

United States (and lately directly by the UN). The Cambodian people for years have been caught in the middle of this international cross fire; with horrendous consequences.

This kind of war is certainly fought "for the hearts and minds" of the people, as everyone professes to know. But what is known and what is done are very different things, as the United States was to learn to its pain in the jungles of Asia.

Vietnam: The First Postmodern War

In response to the strategy of people's war, the United States offered hubris and technoscience. Henry Kissinger proclaimed, "A scientific revolution has, for all practical purposes, removed technical limits from the exercise of power in foreign policy" (quoted in Gibson, 1986, p. 15). Kissinger went from this assumption of unrestrained power to develop his theories of limited war, which he later applied directly to Vietnam. From this first principle of the triumph of scientific reason comes a natural corollary: the supremacy of rationality. David Halberstam noted this in his analysis, *The Best and the Brightest* (1972), which dealt with those rationalists who involved the United States in Vietnam:

> If there was anything that bound the men . . . together, it was the belief that sheer intelligence and rationality could answer and solve anything. . . . [McGeorge] Bundy was a man of applied intelligence, a man who would not land us in trouble by passion and emotion.
> He [McNamara], not only believed in rationality . . . he loved it. It was his only passion. (pp. 57, 288)

> Who is man? Is he a rational animal? If he is, then the goals can ultimately be achieved; if he is not, then there is little point in making the effort. (Robert Strange McNamara, May 1966 [quoted in Kaplan, 1983, p. 337])

> McNamara was coldly clinical, abrupt, almost brutally determined to keep emotional influences out of the inputs and cognitive processes that determined his judgments and decisions. It was only natural, then, that when Robert S. McNamara met the RAND Corporation, the effect was like love at first sight. (Halberstam, 1972, p. 251)

Donna Haraway has called this weird matrix of feelings, "the emotion of no emotion" (personal communication). Power, technology, rationality—all are linked in one of the more influential specific strategies on how to win the war: destroy premodern Vietnam. Harvard professor Samuel Huntington described his policy of strategic hamlets that removed tens of thousands of peasants from ancestral homes as "forced-draft urbanization and modernization" (quoted in Sheehan, 1988, p. 712).

McNamara, the number cruncher from World War II, become President John F. Kennedy's secretary of defense. He hired Charles Hitch, the analyst of firebombing, and latter economics division director at RAND and president of the Operations Research Society of America, to be the Pentagon comptroller. Hitch, in turn, hired Alain Enthoven, another former RAND analyst, to be the deputy assistant secretary of defense for systems analysis, a new position. Other RAND analysts hired by the Pentagon included Harry Rowen as deputy assistant secretary of defense for international security affairs; Frank Trikl, to work for Enthoven on strategic offensive forces (replaced by Fred Hoffman in 1964); and Bill Kaufman and Daniel Ellsberg (a Harvard bargaining theorist like McNamara before him and Kissinger) as consultants (Kaplan, 1983, pp. 252–254).

Fred Kaplan says of Enthoven, "He . . . had the systems analyst's obsessive love for numbers, equations, calculations, along with a certain arrogance that his calculations could reveal truth" (p. 254). Kaplan concludes that these "whiz kids"

> transformed not only the vocabulary and procedural practices of the Pentagon, but also the prevailing philosophy of force and strategy—not only the way that weapons are chosen, but also the way that war should be fought. (p. 256)

One young whiz kid, Barry Bruce-Briggs, enthused:

> Most real military innovation is made over the feelings of the uniformed officers by the so-called whiz kids and defense intellectuals. We've performed this role and for that reason our senior staff people are received at the highest level in the Pentagon like the Jesuit advisers who walked the courts of the Hapsburgs. (quoted in P. Dickson, 1971, p. 93)

This comment reveals a number of things about the whiz kids, including their hubris and affinity for priestly status. There is also something childlike about the assurance of such experts, especially in the light of their paucity of experience and their ignorance of the real cost of war, and considering their dismal successes. Also, such a claim underestimates the commitment many uniformed officers have made to innovation, whether for its utility in war, help in career advancement, or both.

Vietnam was supposed to be fought as a political war, "for the hearts and minds of the people." Instead it became the war of the electronic battlefield, and the United States not only lost the battle for the hearts and minds of the Vietnamese but the struggle for the hearts and minds of many Americans as well. In retrospect, it is clear that the Army's conversion to counterinsurgency doctrine was only superficial (Krepinevich, 1986). The Vietnam War was fought and lost as a conventional high-tech war. Vietnam was run along the assumptions of what has been called crisis management, which is systems analysis (SA) applied to a crisis. Somebody thinks through various possibilities, including the effects of effects (feedback); someone gives key factors certain numerical weights; someone assigns mathematical values (plus, minus, multiply, divide, or something more complex) to the relations between various factors; someone makes assumptions about the value of possible outcomes, and then a machine calculates the costs and benefits of different approaches.

Vietnam was the SA war, the electronic war, the computer war, the technological war. From the point of view of the technophile analysts there was no reason they could have lost—unless it was because the peace movement stabbed the military in the back. Others, including many who were there, saw it differently. The Vietnam War is a particularly good case of how subjugated knowledges can be used to reveal, even change, a society's conception of a complicated issue. Appeals to the wisdom of the grunts (foot soldiers) are made by many writers with quite different explanations as to why the war was lost, but the majority see it as a war that could not have been won.

Obviously, considering their victory, the North Vietnamese knew what the U.S. strategy was. Gen. Vo Nguyen Giap, their commander in chief, described it thus:

> The United States has a strategy based on arithmetic. They question the computers, add and subtract, extract square roots, and then go into action. But arithmetical strategy doesn't work here. If it did, they'd already have exterminated us with their planes. (1970, p. 329)

That even the Vietnamese military leadership could get its views published in New York in 1970 at the height of the war shows how normally discredited views can come out.

One book that has applied many of these marginalized views is James Gibson's *The Perfect War: Technowar in Vietnam* (1986). It is a detailed examination of the Vietnam War as seen by the subjugated knowledges of combat soldiers from the lower ranks, protesters to the war, dissident officials within government, and journalists. It details how official power, patriotic ideology, flashy-destructive technology, and scientific rules of discourse were used to formulate and perpetuate U.S. policy.

Gibson's conclusion is that the system of discourse for Vietnam is the discourse of technowar, an "organized scientific discourse" in Foucault's sense, and that it is sharply regulated:

> Technowar thus monopolized "organized scientific discourse" through multiple, but centralizing relationships among high-bureaucratic position, technobureaucratic or production logic in the structure of its propositions, and the conventional educated prose style. The debate on Vietnam occurs within this unity. (1986, p. 467)

He contrasts the official "unpoetic poetic" of the "technobureaucratic or production logic" (obvious by its propositions and style) with what he calls the "warrior's knowledge," which has many different viewpoints and insights and lacks a formal structure or concepts or any data in a regular sequence.

This warrior's knowledge often comes in the form of stories. The official discourse does not consider stories, poems, memoirs, interviews, and music as valid forms of knowledge. They are disqualified because of their genre and their speaker. Nonfiction is valued over fiction and the high-ranking officer's memoirs are more important than any grunt's:

> The warrior's knowledge is not homogeneous; its insights and concepts and "supporting data" are not laid out in readily understood sequence, but are instead embedded in thousands of stories. . . . Regardless of propositional content, the story form marks the warrior's knowledge as marginal within the terrain of serious discourse. . . . Prose style in the top-level, generalized works of Technowar follows the normal academic practice as well. It is a prose style with a very unpoetic poetic. (p. 468)

Gibson's analysis is that the Vietnam War was lost because it was prosecuted as a rationally managed production system more interested in the appearance of scientificity (body counts, systems analysis) than real effectiveness. He also points out that

> By adopting microeconomics, game theory, systems analysis, and other managerial techniques, the Kennedy administration advanced "limited" war to greater specificity, making it seem much more controllable, manageable, and therefore *desirable* as foreign policy. (p. 80; emphasis added)

McNamara insisted on numbers to explain the war. Many of the numbers were lies. Gibson analyzes a series of reports and other documents, such as the "point system" combat units used to judge their effectiveness, and concludes, after a close reading of Gen. William C. Westmoreland's April 1967 report to President Lyndon B. Johnson, that it presents Technowar as a production system that can be rationally managed and warfare as a kind of activity that can be scientifically determined by constructing computer models. Increase their resources and the war-managers claim to know what will happen. What constitutes their knowledge is an array of numbers—numbers of U.S. and allied forces, numbers of VC [Vietcong] and NVA [North Vietnamese Army] forces, body counts, kill ratios—numbers that appear scientific. (pp. 155–156)

There is also evidence here for theories like that of Robert Jay Lifton (1970, 1987), which argues that what is rationally repressed, the emotional, can return "in crazy or criminal acts." Consider these comments by William Bundy, Robert McNamara, John McNaughton, and Richard Helms about the rationale for the bombing of North Vietnam:

the resumption of bombing after a pause would be even more painful to the population of North Vietnam than a fairly steady rate of bombing. . . .

. . . "water-drip" technique . . .

It is important not to "kill the hostage" by destroying the North Vietnamese assets inside the "Hanoi donut". . . .

Fast/full squeeze . . . progressive squeeze-and-talk . . .

. . . the "hot-cold" treatment . . . the objective of "persuading" Hanoi, which would dictate a program of painful surgical strikes . . .

. . . our "salami-slice" bombing program . . .

. . . ratchet . . .

. . . one more turn of the screw . . .

(all quotes from Ellsberg, 1972, p. 304)

Daniel Ellsberg admits that he heard such talk all the time from these men and, while he often disagreed with the policies they advocated, he never saw what his wife did when she first read them: "It is *the language of torturers*" (304; emphasis added). Torture was the policy:

By early 1965, McNamara's Vietnam strategy was essentially a conventional-war version of the counterforce/no-cities theory—using force as an instrument of coercion, withholding a larger force that could kill the hostage of the enemy's cities if he didn't back down. (Kaplan, 1983, p. 329)

This strategy was based directly on Thomas Schelling's elaboration of game theory in the case of non-zero-sum games, which are those contests where there isn't one winner and one loser, but there's a chance to have a mix of winning and losing (Kaplan, p. 331). In his book *Arms and Influence,* Schelling applies his theory to limited war:

> The power to hurt can be counted among the most impressive attributes of military force. . . . War is always a bargaining process. . . . The bargaining power . . . comes from capacity to hurt, [to cause] sheer pain and damage. (quoted in Kaplan, p. 332)

Or, as Henry Kissinger put it,

> In a limited war the problem is to apply graduated amounts of destruction for limited objectives and also to permit the necessary breathing spaces for political contacts. (quoted in Gibson, 1986, p. 22)

By 1970 it is thus:

> While troops are being brought home, the air war increases. It is a new form of war where machines do most of the killing and destruction. . . . The mechanized war consists of aircraft, huge air bases, and aircraft carriers. The goal of the mechanized war is to replace U.S. personnel with machines. (Crystal, 1982, p. 24)

More than 3 million sorties (a sortie defined as one mission or attack by one plane) were flown by U.S. aircraft during the Vietnam War. Over 1,700 planes were lost, including drones. Over 200 airmen were taken prisoner, and they became some of North Vietnam's strongest bargaining chips. The U.S. Air Force and U.S. Navy ran an ongoing contest to see who could fly the most sorties because much of their budget was determined in that manner. Often planes flew half full or on useless raids just to keep the numbers up. Even though this massive application of air power proved a total failure, some military officers still feel more bombing could have won the war. Their faith in technology is all the stronger after its failure.[3] The various seductions of strategic bombing are more potent in the discourse than any balanced judgment of its efficacy.

Another example of the strange redirections the emotions of war took in Vietnam was the official approval given killing by machines, while killing by people directly was often considered an atrocity: "It was wrong for infantrymen to destroy a village with white-phosphorus grenades, but right for a fighter pilot to drop napalm on it." Civilians could be killed by airplanes but not people: "Ethics seemed to be a matter of distance and technology. You could never go wrong if you killed people at long range with sophisticated weapons" (Caputo, 1977, p. 218).

In *War Without End,* Michael Klare (1972) gives a detailed account of this mechanized war and many of the institutions behind it. He examines the role of human factors research, ergonomics, social systems engineering, modeling, and simulations in the U.S. military's attempts to develop an automated electronic battlefield in Southeast Asia. Along with the scientific management, computerized communications, and data management, this battlefield also depended on sensors for metal, heat, and smells. Collected data was evaluated and sent to Udam, Thailand, where a computer system sorted it out and dispatched hunter-killer teams of helicopters. Anything living in the free-fire zones was a target.

Paul Dickson explains the importance of computers in his detailed history of this system, *The Electronic Battlefield*. One example was the bombing campaign called Igloo White:

> Due to the large number of sensors, the information from them relayed to the Center had to first be digested and sorted by computer before it could be passed along to target analysts who, in turn, passed their assessments to the bases, which control the strike aircraft and order them to their targets. . . . Commonly, the pilot of the F-4 Phantom or whatever would not only not see his target but not even push the button that dropped the bombs—like so much else in Igloo White this was automated with the bombs released at the moment selected by the computer. (1976, p. 85)

This war also saw the first massive use of herbicides like Agent Orange for the destruction not only of rain forests and wetlands but also of agricultural areas, supposedly to improve visibility in the kill zones (Klare, 1972, pp. 169–205). Klare also points out that there was a gigantic infrastructure of researchers who planned these attacks on nature as well as the automation of the Vietnam War. Specifically he mentions RAND, the Special Operations Research Office, the Research Analysis Corp., the Human Resources Research Office (HUMRRO), the Center for Research in Social Systems, the Institute for Defense Analysis (IDA), and the Stanford Research Institute, among others (pp. 80–90). It was IDA, the Joint Chiefs' own think tank, that gave the biggest push to the electronic battlefield. An elite team of moonlighting university scientists called the Jasons proposed an "electronic" wall across the Ho Chi Minh Trail. Although parts were built, it never even slowed the resupply of the North Vietnamese troops in the south.

Another military think tank that played a big role in Vietnam was the Army's HUMRRO, set up at George Washington University in 1951 to research "psychotechnology." It has sought to apply its vision of "human engineering," "human quality control," and the "man/weapon" system to the "mind-bending mission of getting the human weapon to work" (quoted in P. Dickson, 1976, p. 149). HUMRRO had great success:

HUMRRO's influence has been deep and fundamental. It has been the major catalyst in changing traditional training and task assignment procedures from those in effect during World War II to new ones dictated by "the systems-orientation," or training geared to the system that a man is to be part of, whether it be a "rifle system," a helicopter or a missile battery. The HUMRRO approach—and subsequently the Army's—is to look at men as integral parts of a weapons system with specific missions. (p. 149)

James Gibson concurs that this is the military's dominant value system. He claims that

Military strategy becomes a one-factor question about technical forces; success or failure is measured quantitatively. Machine-system meets machine-system and the largest, fastest, most technologically advanced system will win. Any other outcome becomes *unthinkable*. Such is the logic of *Technowar*. . . . Vietnam represents the perfect functioning of this closed, self-referential universe. Vietnam was *The Perfect War*. (1986, pp. 23, 27; emphasis added)

Such was the theory. Details of the practice show both its ideological content and its practical limitations. Heavy investment in computers and automation by the military continued all through this period, with several purposes, including controlling domestic labor:

Automation would "rationalize" the economy by strengthening the control of management, especially the management of big firms. Techniques developed during the war, primarily in the defense industries to compensate for the wartime lack of skilled labor, would be extended to create the "factory of the future," where less and less labor would be needed. Industry could thus replace manpower with machine power and so discipline an increasing obstreperous work force. As Charles Wilson of General Motors, vice chairman of the War Production Board and later secretary of defense, put it in 1949, America had two major problems: "Russia abroad, labor at home." (Roszak, 1986, p. 178)

Autonomous weapons and sensors were to replace manpower on the battlefield. Prototype sensors were deployed right after the Korean War and first saw major combat in Vietnam, as did the first working autonomous weapons. While both have been touted by some as great successes, it is important to note that the United States lost the Vietman War, and it lost most of the battles where they were deployed. The U.S. military abandoned McNamara's wall, which failed to stop infiltration, fought to a bloody draw at Khe Sanh, and suffered total tactical and strategic surprise during the Tet offensive.

The story of the smart bombs and remote-controlled drones is no better, although the latter performed over 2,500 sorties (Canan, 1975, p. 310). So many drones were tried out over North Vietnam that U.S. pilots called it the "Tonkin Gulf Test Range." During the planning for the Son Tay rescue mission, called Operation Polar Circle (the name was chosen by a computer), all seven Buffalo Hunter reconnaissance drones sent failed to discover the camp was empty, and six were shot down (Gabriel, 1985, p. 58).

Smart bombs also had their share of failures. The Falcon, produced at a cost of $2 billion, was effective about 7 percent of the time instead of the 99 percent predicted by tests. Most pilots refused to carry it (Fallows, 1982, p. 55). The Maverick was also a failure, in part due to it being color blind (Coates and Kilian, 1984, p. 155). Ironically, even the one big success of smart bombs, the use of a Hobo bomb to take out the Thanh Hoa Bridge after a number of regular bomb runs failed, was a military failure since the use of a ford nearby meant the North Vietnamese lost little, if any, supply capability.

As the war was running down it became public that these weapons and sensors had been developed without any congressional approval in what was one of the largest and most secret U.S. military research programs ever (P. Dickson, 1976). In scale it was quite comparable to the Manhattan Project and to the gigantic Black Budget research projects of the late 1980s and early 1990s.

Despite the manifest failures and the funding scandal, the research continued and led directly to the present plans for further computerization. On July 13, 1970, General Westmoreland made this prediction to Congress:

> On the battlefield of the future, enemy forces will be located, tracked, and targeted almost instantaneously through the use of data links, computer assisted intelligence evaluation, and automated fire control. . . . I am confident that the American people expect this country to take full advantage of this technology—to welcome and applaud the developments that will replace wherever possible the man with the machine. (Westmoreland, 1969, p. 222)

Or, left unsaid by Westmoreland, the option to make of the man a machine. One of the little-known aspects of the Vietnam War was the wide use of stimulants and other drugs to help elite soldiers perform. Research on a wide variety of compounds for everything from controlling fear to improving night vision was supported by the Pentagon (Manzione, 1986, pp. 36–38).

In his history, *Command in War*, Martin Van Creveld (1985) concludes that the automated and electronic battlefield will be as confusing and chaotic as Vietnam was. On Vietnam he adds "We have seen the future and it does not work" (quoted in Bond, 1987, p. 129). Daniel Ellsberg is a little more brutal: "Whether as field tactic or foreign policy, our way of war now relies on the use of indiscriminate American artillery and airpower that generates innumerable My Lai's as a norm, not as a shocking exceptional case" (1972, p. 236).

They are both right and wrong. Van Creveld is certainly right as far as low-intensity conflicts (LICs) go; but the same high-tech strategy can work remarkably against the right enemy, as Saddam Hussein went out of his way to prove. And Ellsberg is imprecise when he calls the firepower approach "indiscriminate." It was discriminate in the Vietnam War, through numerous restrictions and zones and moratoriums, and it was certainly discriminate in the Gulf War in a very similar way.

Both wars had free-fire zones, especially for the B-52s; they had precision bombing of numerous targets, some of which were horrible misses; they had various targets completely off-limits; they aimed at command, control, communications, and industries in the rear areas and military units in the forward areas; and they were not really challenged in the air. Often targets were prioritized, even chosen, as much for political as for military reasons. Good examples of such targets were the Scud missile launchers and the Iraqi planes bombing Shiite rebels and Kurdish refugee camps.

Since Vietnam, the U.S. infatuation with high-tech war has only increased. Even in the form of war that is the most political in every way, the seductive appeal of war technologies has made technoscience at the center of LIC doctrine. It has developed into a set of interconnected systems, the subject of the next chapter.

Chapter Nine

The Systems
of Postmodern War

New types of warfare do not eliminate older and even
primitive forms.
—Janowitz (1971, p. 417)

Total war itself is surpassed, towards a form of peace more
terrifying still.
—Deleuze and Guattari (1986, p. 119)

The Paradoxes and Structure of Postmodern War

While planning for apocalypse is continual, on the level of actual combat
postmodern war so far has been characterized by limited wars, destabiliza-
tions, revolutions, and the odd midintensity conflict. "Warfare has probably
reached a period of major change," historian David Chandler says for,
"Conventional wars . . . will concede primacy in the spectrum of warfare to
guerrilla and revolutionary struggles in which the political and psychological
factors predominate over the military" (1974, p. 20). But the evidence
actually suggests that, despite the lip service given to the importance of
politics, rhetoric about "political and psychological factors" has often been
followed by policy based on technoscience and illusions about managing war.
If anything, it is war that has replaced politics, not politics war. The possibility
of total apocalypse is always waiting in the wings, after all. Even in the realm
of peacemaking, it is assumed that military force is required. Postmodern wars
take place between two supposed impossibilities: apocalyptic total war and
utopian peace.

On the one hand, apocalyptic war must be prepared for, in order to
prevent it, which makes it inevitable. On the other, this danger of war raging
out of control empowers a militant peace effort, that can become, for some
of the military, a surrogate, the moral equivalent of war. Meanwhile, in the

vacuum, wars continue—cobbled together of various fragments of older discourses from modern, ancient, and even ritual war. This makes for a confusing situation.

The one quality that every postmodern theorist will ascribe to postmodernism is fragmentation. This is the obvious surface "structure" of postmodern war, although there seem to be patterns hidden within the chaos: one is the thread of real war; the other, "the simulation of war and peace." The coexistence of these two different dynamics generates a whole spectrum of contradictions and tensions. From the CBN war(s) that can never be waged (but what's to stop them?) to the extreme technophilia of computer war versus the appropriate technology of people's war and on to lower intensity conflicts that are purely rhetorical and economic.

This period cannot last very long. Statistically some sort of apocalypse is likely, as proliferation continues. But on the ground, where real war is happening peace keeps breaking out. So what are the paradoxes that come out of this? Among the most interesting are the following:

- The main moral justification for war is now peace.
- The main practical justification for repression is the fight for freedom.
- Security comes from putting the very future of the planet in grave risk.
- People are too fragile for the new levels of lethality; machines are too stupid for the complexity of battle. War is becoming cyborgian.
- There is a continual tension between bodies and machines. In purely military terms, machines such as tanks, planes, ships, missiles, and guns are more important than people. But in many countries the human soldiers are much more valuable politically.
- The pace of battle is set by the machines, but it is experienced by the humans.
- Advanced weapon systems are neither machines nor humans, but both: cyborgs.
- The battlefield is really a battlespace. It is now three-dimensional and ranges beyond the atmosphere. It is on thousands of electronic wavelengths. It is on the "homefront" as much as the battlefront.
- The battlespace is also often very constrained. Many targets may not be attacked. Even in war zones the full fury of postmodern weapons is reserved for special killing boxes, free-fire zones, politically acceptable targets, and the actual battleline, if any.
- Battle now is beyond human scale—it is as fast as laser beams; it goes 24 hours a day. It ranges through the frequency spectrum from ultralow to ultrahigh, and it also extends over thousands of miles.
- Politics are so militarized that every act of war needs political preparation and justification. There is only the most limited war space where all important decisions are made on military grounds. Wars can

only be won politically. Through military means the best that can be accomplished is not to lose.

- Obvious genocide, now that it is technologically easy, is morally impossible, for most people—though clearly not all.
- The industrialized countries want colonialism without responsibility (neo); they want empire without casualties.
- Some people in the nonindustrial and industrializing regions want western technology without Western culture; others want both; others want neither.
- Soldiers are no longer uniform. They range from the DoD officials in suits to the women doctors at the front lines, with spies, flacks, analysts, commando-warriors, techs, grunts, desk jockeys, and the like.
- The traditional "male" gender of warriors (soldiers, sailors, and flyers) is collapsing, although it was never absolute. Women can now serve in almost all the subcategories of postmodern warfare except for those dedicated directly to killing (and this is changing as regards combat pilots).
- Civilians, and nature itself, are usually more threatened in battle than the soldiers are.
- New styles of war are invented but old styles of war continue.
- With information war (aka "cyberwar" or "infowar") the most advanced practitioners are the most vulnerable.
- War itself proliferates into the general culture.

All these contradictions stem from the central problem of postmodern war—war itself. Unless war changes radically it will be impossible for war and humanity to coexist. So the old and conservative discourse of war has become wildly experimental and it has institutionalized innovation to an amazing degree. This process has included the colonization of much of Western science and technology as the war system keeps seeking ways to keep war viable. If weapons are incredibly powerful, make them smart. If combat is unbearably horrible for soldiers, make of them machines or make machines soldiers. If war cannot become total, expand it into new realms, war on your own public like the Argentine and Chilean military did in their "dirty" wars or, as is happening in the United States, with the war on drugs (and dealers and users). Practice bloodless inforwar; make business war. Most horrifically, always be ready to destroy the world. If war is impossible, if peace seems to make sense, make ready for the most impossible war—nuclear.

But what gives the system its coherence? It seems that that coherence is not structural but rhetorical. Seen as a discourse system, it's clear that certain key ideas, called tropes in rhetorical analysis, hold the system of postmodern war together.

The Tropes of Postmodern War

What are the primary characteristics of postmodern war? Even in the latest fad, cyberwar, the same basic metaphors are used with the same assumptions that motivated deterrence doctrine, counterinsurgency, and low-intensity conflict (LIC).

Increasing Battlefields: Lethality, Speed, Scope

The nuclear weapon is an example of this, but it should not be considered unique. Biological and chemical weapons can easily equal nuclear devices in lethality and dispersal. The distinction between nuclear and conventional weapons is also dissolving as technoscience works its magic on explosives and their delivery systems. Small clusters of "ordinary" conventional bombs (with new explosives) or single FAE (fuel air explosives) bombs can equal the destructive power of low-yield nuclear weapons. But they represent just a small part of the increased lethality of the postmodern battlefield that will have a killing zone extending 60 miles or more from any front in any direction, deep into space, and at any time of night and day (thanks to infrared and other artificial vision technologies). Because of the incredible growth in rates of fire, explosive force, and delivery system range, the deadliness of infantry weapons has expanded incredibly in the last 40 years.

Dr. Richard Gabriel estimates that war is at least six times more lethal in terms of firepower than it was in World War II. His estimates may be exaggerated, for they come from the military and the weapon makers, but the general thrust of his claim is beyond denying:

> The explosive capacity of most modern weapons exceeds that of World War II weapons by at least five times. Rates of fire have increased by almost ten times. Accuracy has increased by twenty times and the ability to detect enemy targets has increased several hundred percent. . . . The overall result is that the ability of modern armies to deliver a combat punch has increased by at least 600 percent since the end of World War II. Military technology has reached a point where "conventional weapons have unconventional effects." In both conventional war and nuclear war, combatants can no longer be reasonably expected to survive. (1987, p. 153)

These weapons are not only more powerful than in World War II, but they are much quicker now as well. The line between the quick and the dead is often a line between the machine and the human. Virilio and Lotringer argue that speed is central to war today, especially this contradiction between the speed of war and the unquickness of the human body: "There is a struggle . . . between metabolic speed, the speed of the living, and technological speed, the speed of death . . . " (1983, p. 140). Speed also is a matter of distance. The greater the speed, the greater the war measured in time to space.

A LIC battlefield is a country or region. The nuclear battlefield is the world. Netwar takes place in cyberspace. All extend into space, cyberwar most of all. U.S. troops are now expected to fight at night, to fight in the Arctic, to fight in real and simulated space. The battlefield is fragmented in reality and in the minds of the warriors and managers. Getting funding is as much a part of war preparation as planning and training. Protecting the military–industrial complex is as much the armed forces' goal as protecting their countries, if not more.

The Strategy of High Technology

> Technology is America's manifest destiny.
> —*Possony and Pournelle (1970, p. xxxii)*

To try and deal with the postmodern changes in battle conditions, some armies choose old technology in quantity (the Chinese to some extent), others try for current technologies, and the U.S. military has chosen cutting-edge high technology. The U.S. military hopes to solve its problems (desires, missions, demands from politicians and the public) through the deployment of the very best high technologies and formal systems, both dependent on computerization. The U.S. military is explicitly committed to the assumption that it can only achieve battlefield superiority through high-tech weaponry. The militarization of space, the growing use of unmanned and semiautomatic aircraft, the automation of war on and below the sea, and the popularity of cyberwar concepts all show this.

U.S. strategic doctrine has been predicated on technological superiority since World War II. The same principles are being applied today. The strategy for general (so-called conventional and limited nuclear) war for the U.S. Army is based on AirLand Battle doctrine. The year before it became official, the Army Science Board reported that to make the AirLand Battle strategy possible it would be necessary to use high technologies in which the United States has a strong lead. An Army research team stated in 1984: "Future fighting concepts such as AirLand Battle 2000 . . . are largely based on technology" (U.S. Army, 1984, p. 20).

Specifically, the Army has picked five technoscience areas to concentrate on, four of which are directly AI (artificial intelligence) related: "very intelligent surveillance and target acquisition" (VISTA) systems, distributed command, control, communications, and intelligence (DC^3I), self-contained munitions, the soldier–machine interface, and biotechnology (Lindberg, 1984).

The U.S. military is trying to use high technology to cope with "facets of the modern battlefield environment that will be deep, dirty, diffuse, and dynamic" (Gomez and Van Atta, 1984, p. 17). Only with technology, it is thought, can it meet the demands of battle for almost instantaneous actions and reactions, for high-quality intelligence and quick decision making in the midst of a chaotic multidimensional environment, for systems that can withstand high-tech weapons (including biological, chemical, and tactical nuclear), and to maintain military coherence in the face of what are expected to be extraordinary casualty levels ranging up to and including 100 percent dead in many units.

So infatuated with technology have certain parts of the U.S. military become that they even claim that not only will mastery of technology determine victory (instead of mastery of oneself and then the enemy) but that technology is actually driving tactics today:

> Tactical incisiveness depends—and will depend—largely on our mastery in the application of the technology available to us and to our enemies. At present, technology is outstripping the military imagination so swiftly that available hardware will continue to define tactics for a long time. (Peters, 1987, p. 37)

Not only is high technology considered the key element of military tactics and strategy in the United States, its importance is also central to Russian doctrine and the military policy of many other countries, and not just industrialized countries either. Many undeveloped countries have made high-tech weapons their country's major social investment.

The Strategy of Political War

The reverse of the high-tech strategy is to make your military target a political victory. Gilles Deleuze and Félix Guattari call this "guerrilla warfare, minority warfare, revolutionary and popular war" and note that, while war is necessary in this strategy, it is only necessary as a supplement to some other project. Practitioners of political war "can make war only on the condition that they simultaneously create something else, if only new unorganic social relations" (1986, p. 121; emphasis in original). This is, after all, a very old form of war, dating back to prehistory. It contains many elements of ritual war, especially those that were borrowed from the hunt: stalking, hiding, waiting, deceiving, ambushing.

However, what makes this type of war particularly important in postmodern war is that it can defeat the strategy of high technology, as it did in Vietnam. The very same communications and computer advances that make cybernetic high technologies a basis for U.S. strategy today have removed the temporal, geographic, and perhaps even emotional distance that politically allowed neocolonial armies to utilize a genocidal and racist physics- and chemistry-dominated "modern" version of the contemporary high-tech strategy. High technology can still be used to make brief colonial wars publicly palatable, witness the Persian Gulf campaigns, but extended genocidal wars and occupations are now well-nigh impossible politically, nationally, and internationally—unless they are defined as, or declared to be, civil wars (Bosnia, Rwanda, Chechnya, Tibet).

The Changing Soldier

Replacing men with machines has been official policy since World War II. Consider President Eisenhower's New Look military policy with its slogan, "Substitute machines for men!" In 1954, Adm. Arthur W. Radford, Chairman of the Joint Chiefs of Staff, announced that atomic weapons had become "practically conventional" (quoted in Millis, 1956, p. 303). Historian Walter Millis concluded, "The substitution of atomic weapons for uniformed manpower in the ground battle was eagerly accepted on all sides" (p. 314).

Millis noted another implication of the replacement of men by machines that was clear even in World War II: fewer and fewer men were doing the fighting (the teeth), backed by a growing logistical system (the tail):

> Certainly, great numbers [of soldiers] were killed and maimed, yet most of the critical actions seemed to involve relatively few combat men. Thus, the Marines and naval forces who took Tarawa . . . suffered 17 per cent casualties—but there were only 18,000 engaged, out of the 15,000,000 or so whom the United States was to put into uniform. And Tarawa was the decisive action which most clearly foretold the end of the Japanese island empire. Where former wars had represented great clashes of men on extended fronts, here great numbers of men always seemed to be waiting on the sidelines, in support or training or the enormously swollen logistic services, while results of the greatest strategic consequence were achieved by relative handfuls—the famous "few" fighter pilots who defended Britain in 1940, the few who actually flew in the strategic bombers, who waded ashore in invasions, who manned the tanks at the spearheads of the armored divisions. This was the curious result of the introduction of the machine, not simply into war, but into every phase of combat. (p. 258)

Indeed, fewer soldiers were killed in World War II than in World War I. Millions of civilians died in the World War II, however. In the Vietnam War, the United States lost tens of thousands of soldiers, the Vietnamese armies lost hundreds of thousands of soldiers, and the Vietnamese nation lost several million people. In a nuclear war the percentage of military to civilian casualties will obviously skew still further.

Another implication of this is that actual fighting falls more and more to a strange mix of technical and special forces warriors. Jim Stewart notes in an analysis of the U.S. officer corps that "In the past, as many as 75 percent of all officers could reasonably expect to face the enemy with their troops. Now that number has dwindled to a handful of technical warriors" (1988, pp. 19–20). These "technical warriors" are different from the modern soldier in several significant respects, especially their gendering and their relationship to machines in general and weapon systems in particular. Some claim that technical warriors are genderless. As one female U.S. soldier in Saudi Arabia put it: "There aren't any men or women here, just soldiers." But it is more complicated. There is a sharp gender (as status) division between officers and enlisted personnel. Gloria Emerson, the journalist, was talking once to an enlisted man in Vietnam, "a mountain boy from North Carolina." He said that he thought officers "didn't really like women or want them around much." Besides, he added, "We are their women. They've got us" (1985, p. 7). Like subordinates, enemy is often marked as feminine as well.

One of the basic tropes of war is the hatred for the feminine. It's power should not be underestimated. Robin Morgan (1989) shows how this even operated among New Left guerrilla cells in the United States and Europe. Klaus Theweleit's *Male Fantasies* (1989) is an amazingly detailed exploration of this theme. However, in postmodern war there seems to be some shifting of this anger.

As soldiers become more like cyborgs, their gender identity becomes blurred. Cyborgs in general can be either masculine or feminine, although they are often more cyborg than either. Military cyborgs, on the other hand, are still pretty masculine. Since soldiers are also techs, the new masculine identity of soldiers is around mechanization, fixing machines, and working with machines, instead of the traditional masculine identity of physical force, easy access to violence, and the direct subjugation of other men and all women. Women had a hard time fitting into the old masculine category, but the "new male" version is easier to adapt to. It seems the female soldier's identity is beginning to collapse into the archetype soldier persona, creating a basically male, vaguely female, vaguely mechanical image.

The Erasure of Nature as a Category

Nature used to be a force that battles had to be planned around. While this is still true in part, witness the slowdown of air strikes caused by weather during Desert Storm, now weather is seen as a weapon. Perhaps this started with the firestorms of World War II. In Vietnam there were extensive efforts over the course of seven years to use cloud-seeding to cause flooding on the Ho Chi Minh trail (Shapley, 1974).

Just as weather becomes a weapon, terrain is seen as plastic. Thanks to mines especially, the very earth or sea can become hostile. More U.S. soldiers were lost to booby traps and mines in Vietnam than to enemy fire. It is not surprising that some veterans returned hating the very land of Vietnam. In general, postmodern war is not an integration with nature, as ritual war was, or an adaption to natural circumstances, as was necessary in ancient war. It isn't even the attempt to ignore nature, as many practitioners of modern war tried to do, most famously Napoleon and Hitler. In postmodern war nature becomes dominated enough to become another weapon, as with the Hamburg and Tokyo firestorms. Biological, chemical, and nuclear weapons are other examples of turning nature (biology, chemistry, physics) into weapons, but for erasure you can't beat doctrines of cyberwar, which move most of the action into simulated terrain and human consciousnesses.

As nature becomes weapons, community can become a weapon as well. It is then called bureaucracy.

Bureaucracy

John Kenneth Galbraith called bureaucracy the major cause of the Vietnam War. "It was the result of a long series of steps taken in response to a bureaucratic view of the world." He added that "our problem is essentially one of bureaucratic power, of uncontrolled bureaucratic power which . . . governs in its own interest" (Galbraith, 1969b, p. 15). Some analysts give great weight to the way a bureaucracy makes decisions, often labeled *groupthink* (Janis, 1972). Others trace the intimate connections between technology and bureaucracy. For example, in his cultural analysis of what made the United States go to war in Vietnam, Loren Baritz places a heavy emphasis on technology and the bureaucracy it leads to. "Technology demands rationality in the place of individuality," is his premise. He goes on to deduce that

> When the technological mind is turned to the problems of organizing human activity, the result is bureaucracy. This means that an office is created with a predefined function and then a person is sought who meets the specifications of the office. Standardization, technology, never rely on the talents or inspiration of officeholders for solutions. The result, again in its purest form, is impersonality, procedures rather than on-the-spot intelligence, authoritative regulations, not people with authority. (1985, p. 33)

Computer scientists such as Terry Winograd and Marvin Minsky have pointed out that AI programs bear more of a resemblance to how a bureaucracy thinks than to how any individual thinks. Herbert Simon's work started with models of organizational thinking, and from those he has gone on to create models of AI. Could this be part of the mutual attraction between AI and officialdom? No doubt. But like so much of postmodern war it is not the only explanation. The attractions between the AI dream and the lords of postmodern war are varied and flourishing. The next few chapters are detailed examinations of some of the more significant examples. But before going on to them it would be valuable to look closely at how the U.S. military feels about contemporary war as demonstrated through their rhetoric about the most likely form: low-intensity conflict.

The Rhetoric of Low-Intensity Conflict

The Joint Low-Intensity Conflict Project Final Report (JLIC) itself places great weight on the importance of rhetoric (U.S. Army, 1986). At one point the authors quote a British commando colonel to the effect that words are better weapons than bullets:

> Persuading a man to join you is far cheaper than killing him. Words are far, far less expensive than bullets, let alone shells and bombs. Then, too, by killing him you merely deprive the enemy of one soldier. If he is persuaded to join the government forces, the enemy becomes one less, but the government forces become one more, a gain of plus two. (p. 7-1)

They go so far as to claim that the biggest problem the United States faces is rhetorical! The struggle over certain definitions[1] will decide whether the United States or the terroristic "other" prevails in the LICs of today and the near future. To wage its side of this fight the Pentagon has institutionalized the collection of information and the defining of key terms around low-intensity war (Preface, p. xi).

Still, the real problem seems to be the very discussion of it:

> As Americans we consider democracy to be the best form of government, but it is not always the most efficient. The cumbersome decision-making and consensus building process inherent in a democracy can be too slow to respond to dangers before they become critical. This is especially true for threats that are uncertain or ambiguous. (p. 1-1)

In the face of "uncertain" and "ambiguous" threats, this confusion about war, the military feels compelled (by its own perceptions and arguments clearly) into acting. The threat to democracy of the growing complexity of LIC, and its increasing intractability (because of legislative and other constraints), compels the military to respond in undemocratic ways. Inevitably, too much democracy is one of the problems:

Massive increases in political participation may actually increase instability if the government is overwhelmed by conflicting demands. Democracy
may increase the level of conflict in society without providing adequate
institutions for its resolution—especially given the virtually intractable
problems third world countries face. (p. 11-4)

Democracy may well have to be destroyed in order to save it.
It is a domestic struggle in the end, a battle for the homefront.
Through rhetoric, conflict can be made "natural," and the United States
can be defined as an inevitable "arena for struggle" (p. 3-4). And how is
the struggle carried out in the United States? Rhetorically. In the "arena"
of discourse.

So in the U.S. military's own estimation, discourse is the crucial
domain, and weapon, in postmodern war—a weapon that must be
"forge[d]" out of the various voices of "interests" in the United States.
The problem is partially one of definition, as noted earlier. But it is also
described more candidly. It turns out to be the Vietnam syndrome, and
they name it themselves (p. 3-4).

The military also performs its own discourse analysis of the definition
of LIC as distorted by the Vietnam syndrome:

> The United States is poorly postured institutionally, materially, and
> psychologically for low-intensity conflict. Much of the problem concerns
> the very meaning of the term, which emerged as a euphemism for
> "counterinsurgency" when that term lost favor. . . . Largely as a result of
> Vietnam, the mood of the government and the nation has shifted away
> from wanting to deal with "dirty little wars." . . . [The United States's]
> preoccupation and loss of faith in itself as a result of the Vietnam War
> encouraged our enemies to be bolder and discouraged our friends, who
> came to doubt our wisdom and our reliability. (p. 1-9)

This is not just a general complaint. The report argues that these discourse
rules apply to real policy and hinder its execution. They cite the proinsurgency support for the Contras. One of the main problems with the assistance
for what the Pentagon liked to call the "Nicaraguan resistance to the
Marxist–Leninist Sandinista regime" was public interest:

> Intense scrutiny of political, moral, and practical aspects of United States
> assistance by the media, United States policy makers, and the public has
> resulted in a general inconsistency in the type, amount, and availability
> of United States assistance. (p. 2-2)

But it turns out the heart of this "political" and "moral" problem is technical
("practical") after all; it is the invention and spread of electronic media:

The advent of electronic media has brought the gruesome aspects of political violence into the living rooms of millions of people worldwide. The result has been instant recognition for formerly unknown insurgent and terrorist groups. In addition, media coverage has led to an intense scrutiny of United States policies and actions. (p. 2-2)

It is a theme that this study continually returns to. Technology determines politics. Technoscience is politics by other means. It is related to the problem of "Western conscience" (presumably the "Eastern conscience" doesn't have the same difficulties). Technoscience causes problems, which the Western conscience is bothered by. The thought of doing something about technoscience is never spoken officially; technoscience will solve the problems. But the conscience is pesky, so perhaps it can be minimized. Therefore, from the military perspective it is the "modern news media" that remains the main obstacle:

> Protracted war troubles the Western conscience. It causes people to doubt their purposes and to assume that the enemy may have a just cause. Protracted conflict also increases the ambiguities of the situation, and the modern news media will bring these ambiguities home for public debate, exacerbating the uncertainties and compounding the difficulties of involvement. (p. 4-13)

Manipulating the media therefore becomes a crucial part of any low-intensity military operation:

> The media exerts a powerful, if indeterminate, influence on public opinion, and this can have an impact upon operations, either for good or ill. Political and military leaders must consider the media's role and develop appropriate programs and relationships that will sustain operations. (p. 6-12)

The report goes on to claim the media and the Pentagon should cooperate. But on whose terms?

It turns out that the plan is to use the press as much as possible. Volume II, Low-Intensity Conflict Issues and Recommendations, is classified. In its unclassified table of contents we find, Section H: Public Information and Support, Issue H1: "The Need to Use Media Coverage and the Free Press to Further United States Operational Objectives" (p. ix). Gen. H. Norman Schwarzkopf admitted doing this during the Persian Gulf War.

So, if the main issues are rhetorical, what of the central term, LIC, itself?

> Low-intensity conflict is neither war nor peace. It is an improbable compilation of dissimilar phenomena that, like the Cheshire cat—which

seems to fade in and out as you look at it, leaving only its mocking smile—bedevils efforts at comprehension. (p. 16-1)

> The term "low-intensity" suggests a contrast to mid- or high-intensity conflict—a spectrum of warfare. Low-intensity conflict, however, cannot be understood to mean simply the degree of violence involved. Low-intensity conflict has more to do with the nature of the violence—the strategy that guides it and the way individuals engage each other in it—than with level or numbers. (p. 1-2)

These definitions are both from the same report. This obvious tension, even disagreement, about the meaning and definition of LIC is considered a serious problem: "No single issue has impeded the development of policy, strategy, doctrine, training, or organizations more than the lack of an approved definition of low-intensity conflict" (p. 1-2). Finally, after "years" of work, the Joint Chiefs approved an official definition:

> Low-intensity conflict is a limited politico-military struggle to achieve political, social, economic, or psychological objectives. It is often protracted and ranges from diplomatic, economic, and psychosocial pressures through terrorism and insurgency. Low-intensity conflict is generally confined to a geographic area and is often characterized by constraints on the weaponry, tactics, and the level of violence. (p. 1-2)

Notice how the definition must be formal, like a computer language. The longer definition hashed out by the project is an even greater reflection of this. It includes five major subcategories: Insurgency; Counterinsurgency; Terrorism Counteraction; Peacetime Contingency; and Peacekeeping.

Included within the category of Peacetime Contingency are many violent actions (although they use the euphemism "conditions" that are "short of conventional war." Some of the specifics mentioned are "strike, raid, rescue, recovery, demonstration, show of force, unconventional warfare, and intelligence operations" (p. 1-3; cited from TRADOC Pam. 525-44). These are obviously options usually for the "local" arena, the second crucial battle site:

> The local arena is crucial because of two significant developments: the emergence since World War II of a number of new, independent states whose sociopolitical stability is often fragile; and the growth (in various societies) of groups, often with international connections, that are dedicated to radical change through violent means. (p. 1-4)

To explain this threat, the report uses a strange mix of historical and Manichaean assumptions. First, the problem is the instability caused by the postmodern collapse of colonialism and the resultant crisis of modernization.

Because "all societies have varying degrees of instability at various times" and "modernization and rapid development can seriously undermine traditional values, patterns of organization, and older forms of social cohesion," and since "nation building and rapid economic development are disruptive by nature," there is "the basis for instability and violent change."

But, along with the historical view that "these [are] inherent features of the modern world," coexists the claim that evil is real, in the form of "groups dedicated to bringing about a radical change of power regardless of the sociopolitical conditions of a given society." Why do they want to do this? Remember! According to the DoD, "the presence of legitimate grievances is only an excuse." Do they do it for love? For money? No, for the sake of "terrorism" pure and simple (p. 1-5).

And how is this force, terrorism, defined? "Terrorism consists of a series of carefully planned and ruthlessly executed military-like operations" (p. 5-3). Why is there terrorism?

> Terrorism is carried out purposefully, in a cold-blooded, calculated fashion. The men and women who plan and execute these precision operations are neither crazy nor mad. They are very resourceful and competent criminals, systematically and intelligently attacking legally constituted nations that, for the most part, believe in the protection of individual rights and respect for the law. Nations that use terror to maintain the government are terrorists themselves. (p. 5-3)

At the same time, powerful constituencies cannot be offended. What good are nuclear weapons or even armored divisions against terrorists? The other threats have to be remembered. So, the report notes carefully that

> The Soviets are the great menace and much of United States policy and the raison d'être of United States international involvement derive from this threat. Budgets and programs, the meat and potatoes of influence in government, depend upon it. Furthermore, the United States public's normal penchant for isolationism and self-absorption—its inertia, short of clear, immediate, impending crisis—is overcome only by conjuring with this menace. (p. 1-10)

How honest, and how revealing now that the Soviet Union is no more. Where's the "menace" to come from if the Soviet Union is not a threat? What "reason for existence" (is it less obvious in French?) for the military then? Why would they get their "meat and potatoes" from the taxpayers? Why "conjuring"? Because people still don't see! They don't understand! "No constraint . . . is more powerful than our inability to comprehend the threat that faces us"(p. 1-10) the report proclaims. So the United States must prepare for a world that is perplexing and dangerous. As a superpower in the nuclear age with an economy largely dependent upon an extensive, vulnerable overseas trade system, this country faces challenges that are far more troubling and complicated than those that it faced before World War II. (p. 1-1)

U.S. interests are global. So, by being the dominant world power, the United States is in a "more troubling and complicated" position than before World War II. Because it is the world's dominant trader it is "vulnerable." The better things are, the worse they are. Every revolution is a threat to Pax America:

Insurgency poses an open and direct threat to the ordering of society. . . . For the United States, with extensive global interests and an economy increasingly reliant on a stable world order, the chronic instability in the third world is a serious concern. (p. 4-1)

So even LIC becomes a threat, yet successful peace would mean the end of postmodern war.

Rhetorical Psychodynamics

We live in and through the act of discourse.
—George Steiner

There is no meaning without language games; and no language games without forms of life.
—Ludwig Wittgenstein

Man is an animal suspended in webs of significance he himself has spun.
—Clifford Geertz[2]

This language, this official-technical discourse that is the justification and rationalization for a very important and expensive military and scientific program, is supposed to be logical and free of rhetoric and emotion. Instead, the text is full of emotionally evocative figures of speech and images. Often in the case of technical material, the argument *depends* on this rhetoric or on the flowcharts, graphs, and other scientific tropes used, especially the hypertechnical language of bips, mips, gflops, mflops, and so on.

In many crucial ways, the perspective evident in postmodern war discourse is a case study of what Zoë Sofia (1984), discussed earlier in Chapter 5, has termed "The Big Science Worldview." Some of its key components are as follows:

- *Epistemophilia*—the "obsessive quest for knowledge." It is certainly a common theme in military rhetoric, and it is the basic passion of all hard science fiction and, of course, of all science. While this passion for knowledge is not necessarily always a bad thing, as with all loves it can go wildly out of control.

- *Upward displacement* is most clearly seen in the extraterrestrial longings of the military's desire for militarizing space. It is the sacred high ground. It also is reflected in their boundless faith in "high" technology.

- *Half-lives* are desirable. Types of half-lives include inorganic intelligence (especially artificial), cybernetic organisms, aliens, and other semiliving subjects such as ghosts. Society's fascination with such creatures includes the military and science. Cyborgs are central here.

- *Cosmology recapitulates erogeny* is a fancy way of saying that how people view the world reveals what they think and feel to be erotic. In this light, military futurists, who believe mainly in technoscience and future war, have a similar erotics to the Italian Futurist art movement that wrote odes to the machine gun. F. T. Marinetti proclaimed in the *Futurist Manifesto*:

> We will glorify war—the world's only hygiene—militarism, patriotism, the destructive gesture of freedom-bringers, beautiful ideas worth dying for and scorn for woman. (quoted in Lifton and Humphrey, 1984, p. 60)

- *Logospermatechnos* is that the mind breeds real brainchildren as myths and technologies.

There are other perspectives with their own epistemologies in computer science. While "The Big Science Worldview" is dominant in AI, there is also an alternative theory of knowledge that can be clearly seen in the debate among computer scientists about the viability of the Strategic Computing Program (SCP) and the Strategic Defense Initiative (SDI).

The rules of evidence that count in this debate are quite different from the rigid replicability and experimental claims of the physical sciences. AI is a science of the artificial, not the natural, and especially it is a science of unions with the artificial. A good example of some of the applicable metarules can be found in the technical debate among computer scientists about military AI. It isn't really an empirical debate; rather, it is creative and experiential.

Against the claims of some AI scientists that these military projects can indeed be made, others respond that they represent technological hubris, that their own experiences show that such projects will fail. For an argument the dissenters are as likely to tell a story as to cite a study. Certain stories of interesting computer bugs are used often. Favorites are (1) the rising moon sets off Early Warning System; (2) Apollo and shuttle software failures; (3)

DIVAD (Division Air Defense System) computer-controlled cannon aims at a review panel of VIPs instead of its target drone. The effect of these perspectives can be seen in the official Eastport Study Group report (1985), which admits in the section "On the Nature and Limitations of Software Research" that

> if one wants to decide which software development technique is most appropriate for a particular subset of the battle management software, one can not make an objective assessment; it will rely at least partially on anecdotal evidence and the subjective judgement of experienced people. (1985, pp. 46–47)

This is because with software design "issues of how humans can comprehend notations, express computation in them, and exchange complex, detailed information with one another are paramount." The report points out on the next page that

> assessments of software development techniques have been largely qualitative. Indeed many of the well-known papers in the programming languages and operating systems disciplines have a distinct flavor of literary criticism. (p. 48)

This emphasis on language and interpretation is part of the infomania of postmodernism. Along with mathematics, really a language for many computerists, natural language is at the core of computing, especially AI research. Computer languages, mathematical-logical, are a lingua franca of many a postmodern neighborhood today, especially in the military community.

But this is not just talk that will be understood in any human community. It is the street talk of the neighborhoods of high science, the latest technologies, and postmodern warriors. It is language that shapes policy and policy makers. It is language that yearns for new life and real power from the formal artificial languages that form the computer programs of the SCP. What kind of life does this discourse shape? What webs of significance does it spin around us? What meanings from this particular game and the forms of life it plays with?

The remaining chapters will explore its possible futures, and the forces that will determine which comes into being.

Part Three

The Future

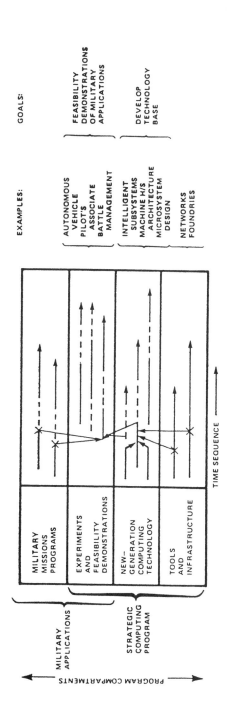

Diagraming the future. These figures are from the original Strategic Computing Report. They map out the imaginary future of military computing. The charts are organized to move up and forward, and are reproduced as giant foldouts in the actual report. They predicted numerous technological and scientific breakthroughs, very few of which came to pass, but they were successful in garnering large amounts of money from the U.S. Congress, which was their real goal after all. In their crude timeline/flowchart progress they recapitulate the progress of many military technoscience projects from the emotional rhetoric of hope and fear at the beginning to the nuts and bolts and millions of dollars of the actual programs as living and growing bureaucracies. Charts from Strategic Computing—A New Generation of Computing Technology, Department of Defense, October 28, 1983.

BE A MAN AND DO IT
UNITED STATES NAVY
RECRUITING STATION
34 East 23rd Street, New York

Be a man and do it. War has been a "man's thing" for millennia, with a few notable exceptions. But as this World War I recruiting poster demonstrates, women have played important roles encouraging the men. War itself is highly sexualized, and often weapon systems themselves sexualized, as these leggy ICBM booster rockets are in a cartoon from the Minuteman Service News *advocating systematic design reviews. Actually, though, despite images such as these, the line between men and women in war is rapidly dissolving. When integrated intimately with machines, mechanical and computer skills are much more important attributes then protruding genitalia. Gender is growing irrelevant for soldier cyborgs. World War I poster by Howard Chandler Christy, Imperial War Museum, London. Cartoon from* Minuteman Service News, *no. 33, July–August 1967, p. 3.*

War is hell, unless you're a 40-foot robot. This Atari ad for the Iron Soldier game illustrates the attraction of postmodern war to the technophilic prepubescent imaginary. Incredible power, bloodless death, mechanized embodiment, multiple lives are very attractive. At least Atari knows it is only a video game. It is not an accident that the military now mobilizes science fiction writers and other futurologists to plan for the wars of tomorrow just as they consciously recruit video-game-playing adolescents to fight the same conflicts. Atari Iron Soldier advertisement from 1995.

Future operations fantasy. This cover from the most important U.S. military journal demonstrates that science fiction can become military policy. When read in detail, the military's plans for cyborg soldiers deploying state-of-the-art and yet-to-be invented military technologies are too disturbing to be amusing. Fear of technological surprise has become a mania for the U.S. military, leading to funding for everything from psychic protection for ICBM silos to research on antigravity rays. As if nuclear weapons weren't enough, various biological, chemical, amplified light, ultrasonic sound, and other weapons are continually being developed, in the quest for the magic bullet that will win all wars. Of course the main force amplifier to capture the military's imagination is information itself. You can be sure that the soldiers, tanks, and air frames in the illustration are linked together and to various headquarters through the magic of command, control, communications, computers, and information systems. But information is not knowledge, nor is it wisdom, and too much of it can easily lead to defeat when it is confused for either, as the wars in Vietnam and Afghanistan demonstrated. "Future Operations" cover from Military Review, *vol. 73, no. 11, November 1993.*

Happy bomber guys. USAF officers watch the B1 for the first time at the Farnborough Air Show in 1982. Are these guys happy or what? It is not our future that they see, but a very beautiful and expensive plane, with a life all of its own. More care is given the future of the B1 by the political powers that be then most cities can command. As long as weapon systems command this loyalty, and produce this kind of pleasure, they will proliferate. Mike Abraham/Network.

Bombers and the sacred mountain. The question is, bombers or the sacred mountain? This World War II image of U.S. B-29s on their way to bomb Japan reflects our current dilemma. We are poised between what we understand all too well, the destruction represented by these bombers, cyborgian war machines probably filled with incendiaries and targeted for working class rice-paper houses, and what we can never explain, the sacred. The sacred can only be experienced and appreciated, not rationalized. Ironically enough, much of the attraction of war is the same. It is sacred by dint of the sacrifice (destruction of life) it calls for, and it can be beauty beyond words, just as Mt. Fujiyama is. But war is certainly not rational now, if it ever was. Stripped of this pretense it is clearly insane. Beautiful at times, certainly, but insane. Whether or not the benefits and costs of war will be truly weighed today will determine the quality, and the very possibility, of our tomorrows. USAAF files, NASM Library, photo 3139.

Chapter Ten

The Cyborg Soldier: Future/Present

> Recall that it is not only that men make wars, but that wars make men.
> —*Ehrenreich (1987b, p. xvi)*

> Cyberwar is about organization as much as technology. It implies new man–machine interfaces that amplify man's capabilities, not a separation of man and machine.
> —*Arquilla and Ronfeldt (1993)*

> At the outset we will assume that the "man–machine" system is more perfect than "man" (people) or "machine alone."
> —*Druzhinin and Kontorov (1975, p. 16)*

Soldiers Future/Present: Popular Images Troops

Wars do make men. And not just real wars. Possible wars, imagined wars, even unthinkable wars shape men—and women. Just as modern war required modern soldiers, postmodern war needs soldiers with new military virtues who can meet the incredible requirements of high-tech war. These new soldiers are molded, in part, by personnel science and marketing analysis in uneasy alliance with traditional military discipline and community. But in another sense it is the weapons themselves that are constructing the U.S. soldier of today and tomorrow.

Weapons have always played an important role in war, from the gear of the Greek hoplite to the tankers of the world wars. Today, however, it is not that the soldier is influenced by the weapons used; now he or she is (re)constructed and (re)programmed to fit integrally into weapon *systems*. The basic currency of war, the human body, is the site of these modifications, whether it is of the "wetware" (the mind and hormones), the "software" (habits, skills, disciplines), or the "hardware" (the physical body). To overcome the limitations of yesterday's soldier, as well as the limitations of automation as such, the military is moving toward a more subtle man–machine integration: a cybernetic organism (cyborg) model of the soldier that combines machine-like endurance with a redefined human intellect subordinated to the overall weapon system.

Current DoD policy is creating a postmodern army of war machines, war managers, and robotized warriors. Logistics command sees the soldier as a digitalized "manprint," as Major General Wallace C. Arnold explains (1995). For him the key issue is the soldier–information interface. The ideal postmodern enlisted soldier is either an actual machine (information processing) or will be made to act like one through the psychotechnologies of drugs, discipline, and management. The ideal postmodern officer is a skilled professional who manages weapon systems and sometimes applies them in combat. In all cases soldiers are to be intimately connected with computers through hard wiring, lasers, and more traditional soldier–machine interfaces.

That is why in North America the cozy often comical images from modern war of U.S. (Kilroys, dogfaces, G.I. Joes) and Commonwealth (wisecracking Tommys, soft-spoken Canucks, loud Anzacs, and silent Gurkhas) soldiers from World War II (*Combat, Hogan's Heroes, McHale's Navy*) have been shattered. In film portrayals of the first postmodern war, Vietnam, multiethnic teams of clean-cut born-in-America guys (*The Green Berets*) coexist with men trapped in an insoluble moral dilemma (*Deerhunter, Platoon*) or even inside of *Apocalypse Now*. The Vietnam War means Vietnam protests (*Coming Home, Fields of Stone*), as even the charmed circle of all-American guys (*Tour of Duty*) must recognize. The first TV show about Vietnam, M*A*S*H, was set in Korea during that conflict, but it was always, psychologically and ideologically, about Vietnam. It was one of the most successful TV shows ever in the United States, and it still plays out its ironic antiwar message every day in syndication. Its dark humor is copied by some (*Good Morning Vietnam*), and its feminine healing focus by others (*China Beach*); but it remains unique, marking as it does in the mass media the critique of modern war first elaborated by the poet veterans of World War I and then by the novelist veterans of World War II.[1]

As the reasons for war have become less clear, the moral standing of the participants has also become very confusing to some of us. "Bad guys" and "good guys" (which translate as "us" and "them" or "friend" and "foe" in the jargon of counterterrorist experts, police, and the military) are harder to keep separate when one compares:

- Stateless terrorists and state terrorists
- "Right Stuff" astronauts ("one giant step for mankind") and "Top

Gun" pilots (killing Mu'ammar Gadhafi's four-year-old daughter with a smart bomb)

- Jedi Knights (modeled on the Vietcong according to filmmaker George Lucas, the creator of *Star Wars*), the nickname given the war planning staff of General Schwarzkopf's Gulf War command (Gordon and Trainor, 1995, p. 126), and Rambo (a Nam vet whose first movie is all about killing policemen)

- Fat Pentagon officers smoking cigars with briefcases full of money (as shown in countless political cartoons), and naive soldier-engineers (Jimmy Carter)

- Losing superpower soldiers riding their tanks and helicopters out of Afghanistan and Vietnam, and the winners

- Good cyborgs (*Robocop, D.A.R.Y.L.*, and hundreds of cartoon and sci-fi characters) and bad cyborgs (*The Terminator* and more hundreds of cartoon and sci-fi characters), often in the same movie (*Terminator II, Robocop II*)

- Elite U.S. soldiers invading countries (Grenada, Panama), training death squads (Latin America), killing civilians with high-tech weapons (Persian Gulf, Libya), and as victims of terrorism (Marines in Beirut, woman soldier in Germany, black sailor on hijacked plane, Scud victims in Saudi Arabia)

- Boring, nerdy war researchers (as characterized in David Broad's book *Star Warriors*) and Dr. Strangelove scientists

G.I. Joe is now a TV character who fights more battles with transformers (robots that can turn into cars and planes) and dinosaurs than any normal human enemies. Almost all of the violent children's cartoons involve cyborgs and other strange mixes of the human, the beastly, the alien, and the technological. Some shows allow the kids at home to join in the fight as well by using expensive interactive toys. In just the last four years of the 1980s the main war cartoons included (with their toy companies in **bold** and their TV network or producer in small capitals: the interactive *Captain Power and the Soldiers of the Future* (**Mattel**); interactive *Photon* (MCA–Universal); interactive *TechForce* (**Axlon**); *Inhumanoids* (**Hasbro**); *Centurians: Power Xtreme* (**Kenner**); *G.I. Joe* (**Hasbro**); *Challenge of the Gobots* (**Tonka**); *Transformers* (**Hasbro**); *Dungeons & Dragons* (**TSR**, CBS); *Rambo: Force of Freedom* (**Coleco**); *Star Wars: Droids* (**Kenner**, ABC); *She-Ra, Princess of Power* (**Mattel**); *Thundercats* (MCA–Universal); *Jaycee and the Wheeled Warriors* (**Mattel**); *Superpowers* (**Marvel Comics**, ABC); *Voltron* (**Matchbox**); *Lazer Tag* (**Worlds of Wonder**, NBC); *Star Wars: Ewoks* (**Kenner**, ABC); *Silverhawks* (**Kenner**); *HeMan & Masters of the Universe* (**Mattel**); *Robotech* (**Matchbox**); *Dinosaucers* and *Dinoriders* (**Legos**).[2]

Postmodern war is as disjointed as the cartoons, even if it does lack dinosaurs. It is both the extrapolation of modern war motifs and weapons into their *reductio ad absurdum* and their unabsurd opposite images. So the atomic bomb and the computer, the two great military brainchildren of World War II, become nuclear overkill facing hijackers armed with handguns and the electronic battlefield is overrun (and undermined) by the agrarian Vietnamese. And, as alluded to above, the common-man foot soldier has assumed any number of other identities from a female soldier-tech repairing a faceless machine to an elite bloody-minded covert action warrior using a satellite to call in a killer droid.

In important ways the spies and spymasters of the CIA and other intelligence agencies have to be considered the precursors, and the comrades, of today's soldiers, especially the elite (counter)terrorist troops and commandos of the Marines and Delta Force. Since the 1950s the CIA has tried to use drugs, electricity, and hypnotism to turn some intelligence agents, their own and the opposition's, into functional robots. As one CIA psychologist put it, "The problem of every intelligence operation is how do you remove the human element" (quoted in Marks, 1979, p. 49). Just exactly how successful these attempts have been is impossible to find out, but several clear trends are obvious:

1. The replacement of most human intelligence collected by agents with signal and satellite data
2. The widespread experimentation on the use of drugs and hypnosis to create amnesia and to facilitate the reprogramming of agents
3. The use of direct electrical implants (bioelectronics) to control behavior

CIA experiments on many animals, including monkeys, dogs, cats, crows, and various reptiles (among others), certainly took place and may well be continuing. A CIA document from April 1961 admits that the CIA had a "production capability" in direct electrical brain simulation and that they were "close to having debugged a prototype system whereby dogs can be guided along specific courses." Less than a year later another CIA report claimed that "the feasibility of remote control of activities in several species of animals has been demonstrated." It went on to promise that "investigations and evaluations" would be aimed at applying "these techniques to man." What the CIA has managed to develop in the almost three decades since these reports is not public knowledge but research continued for at least ten years and possibly to this day (Marks, 1979, pp. 209–211).

War Managers and Trainees

My image in some places is of a monster of some kind who wants to pull a string and manipulate people. Nothing could

> be further from the truth. People are manipulated; I just want
> them to be manipulated more effectively.
> —B. F. Skinner (quoted in Marks, 1979, p. 214)

In many ways, trying to make soldiers into machines has preceded making machines into soldiers. Uniforms, hierarchies, discipline, training, and rules of war have all been used for some time to try to control soldiers, to make them interchangeable, and to mold them into an effective fighting unit. Since the early 1900s modern labor management (capitalist and state social-ist) has coevolved with military personnel management. Both aim at fitting the individual worker-soldier into the military–industrial system.

Lawrence Radine, in his book *The Taming of the Troops: Social Control in the United States Army*, traces how current military managers have added to their traditional forms of "coercive" and "professional paternalist control" a new form he calls "co-optive rational control through behavioral science and management." In his view personal leadership is being replaced by "testing, attitude surveys, various utilitarian and life-style incentives, and weapons systems" (1977, p. 91).

He points out that

> One application of this social engineering approach is the way the military matches men to machines (as well as matching some aspects of machines to men). The man is the extension of such machines as artillery pieces or weapons systems generally; he is an adjunct for some limitation the machine has due to its incomplete development. (p. 89)

An important part of this is the very detailed study of the humans in their system context: "[The soldier's] performance is measured and pre-dicted to a degree of precision unmatched in previous human experi-ence." Measurement of behavior leads to more effective social control and might even "provide a nearly unlimited potential for the bureau-cratic or administrative domination of man." He credits C. Wright Mills for first describing this style of domination, and coining the term "cheer-ful robots" for the type of subjects it produces. Radine concludes that "the ultimate result of co-optive rational controls is cheerful states of mind, with no values or beliefs other than one's own comforts, and automatic, mechanical performance" (p. 90).

The rationalization of social control in the U.S. military has many applications. In training it involves using B. F. Skinner's principles of rein-forcement and punishment as operant conditioning. Survey research is used to identify effective reinforcers, and they are administered carefully to maximize their impact. Just as important is the systems approach.

Radine notes that "social control is more effective when it is embedded in the totality of a situation," and then he reflects upon the importance technology can have in maximizing the control of individual soldiers:

As weaponry gets more complex and is based more on hardware than on manpower, the interaction of various components becomes emphasized and is termed a system or a weapons system. The weapons system becomes, perhaps in an unanticipated way, a new and effective technique of domination. It elicits obedience and makes resistance appear senseless. The other development associated with the modernization of weaponry and hardware is the principle of training men who operate weapons systems through simulating the environment with a computer. (1977, p. 130)

As part of a system the individual soldier has less of a chance to deviate from expected behavior. Realistic simulations also serve as "a means of indoctrination" both because they serve to validate the system itself and because, for a technosoldier launching a missile, simulated conditions and real-war conditions are almost identical. Through systems analysis, social psychology, behavioral sociology, personnel management, and computer-mediated systems, the individual soldiers becomes part of a formal weapons system that is very difficult to resist. It produces "a kind of isolation" from the violence of war that allows for its unrestrained prosecution because the bomber pilots and other distant killers are removed from the bloody results of their decisions. And "the structuring of a situation through the use of technologically developed equipment and realistic team training" produces a "degree of conformity and effectiveness" that is much better than traditional leadership, because it "is very difficult for the individual to sense the degree to which this form of domination can control his behavior" (p. 142).

It is important to remember that as one goes up the chain of command the officers are supposed to be controlling, not controlled. In many cases they certainly shape events, but at times it seems they act as unconsciously as any enlisted soldier. Certain desires seem to dominate some military men. These hidden emotions return in the form of hubris and hatred. Under President Reagan they almost took over U.S. foreign policy, as a tight circle of military officers and spies with control of the National Security Council and the ear of the president sought to effect the military rollback of the communist menace around the world (W. Kennedy, 1987, p. 8; Zakaria, 1987, p. 19). But men like Oliver North, the zealot warrior who *on his own time* returned to Vietnam and went on dangerous combat missions, are the exception. Most officers who rise to power are cooler technocrats with better impulse control, such as the intelligence admirals (Bobby Inman, Stansfield Turner, and John M. Poindexter) or those consummate politicians Gen. Colin Powell and Gen. H. Norman Schwarzkopf. They wanted to manage the Cold War with the Soviets, and they want to manage the current Cool War with the Third World, not fight it. They despise Ollie North because of his political approach to what they see as a technical problem: managing the U.S. empire. Besides, he didn't respect the military's own hierarchy, and he even put Republican politics above the needs of the military as an institution.

But their illusions are the same as North's, by and large: great faith in technology; unreflective belief in the rightness of the American empire; incredible hubris combined with the fear and anger of those whose profession is poised on the brink of exterminating the human race. But they don't get excited. They send a memo, make a proposal, procure that new weapon or management system that will really get the problem under control. And they don't forget their careers.

Thus, most of today's officers are technocrats, not combat leaders. In the Air Force almost 60 percent of the officers have graduate degrees, and even the Army boasts that a quarter of its officers have a Master's or a Ph.D. degree (Satchell, 1988, p. 2). Roughly half the Navy's officers are line officers who might see combat; the other half are in noncombat specialties. There is no combat specialty at all, while three systems of technomanagerial specialization coexist, each with powerful bureaucratic sponsors. While 200 Navy officers were in senior service colleges learning "warrior craft" in 1985 some 1,200 were in full-time pursuit of advanced degrees in science, engineering, and management. Also, while the Navy has over 200 public affairs officers, less than 30 officers teach coordinated battle group tactics. Instead of "the management of violence" there is "management for its own sake" (Byron, 1985, p. 68). Actually the Navy is quite right to worry about public affairs more than coordinated battle group tactics, as cuts in public spending are much more of a threat to the Navy than any enemy fleet. The management of information, even the public meanings of naval battles, is of crucial importance.

Studies of World War II by historian S. L. A. Marshall (1947) purported to show that less than 25 percent of the men in combat were really shooting at the enemy. Some men didn't want to kill, he claimed, while others were too afraid to try. While Marshall's research has been discredited,[3] there is little doubt that war is almost off the human scale. It is also very revealing that the U.S. military establishment was so willing to believe Marshall's theories even though most World War II combat troops have always considered them bunk. Here is a case of a pseudoscientific claim (Marshall said he had conducted thousands of after-combat interviews as the basis for his research) becoming "fact" in military discourse even though it was denied by combat veterans (including generals such as Maxell Taylor and Matthew Ridgeway), and it hardly reflected well on the image of the American fighting man either. The reason Marshall's ideas were so popular is that they paraded as science, they confirmed the belief of many high officials (civilian and military) that war could be studied and managed, and they supported the drive for weapons systems in detail.

Later, more valid psychological studies showed that almost every human has a limit as to how much battle he or she can stand. Only 2 percent of all examined soldiers were capable of continued heavy combat of more than a few months. The vast majority of this 2 percent tested out on standard psychological profiles as pure psychopaths with no conscience or emotional involvement in the killing and dying around them. They can act coldly, with calculation and aggression but not blood lust. Producing more such soldiers seems to be the aim of many training schemes and significant military drug research (Gabriel, 1987).

To the bureaucrats one of the biggest management problems is finding, or making, soldiers like this. To manage war there needs to be a way to facilitate the average soldier performing well under the extraordinary stress of war today. The World War II studies showed that traditional training does not suffice. So, years ago, the Pentagon began its search for ways to improve the integration of human soldiers into the inhuman battlefield, at least until there are more effective killer robots. The DoD will try almost anything.

Training for postmodern war involves a central paradox, however: potential external stimuli (death and destruction) are much greater than in previous wars, but the duties of the soldiers are more technical and complicated. How can psychological limits be overcome and yet human judgment preserved? This was one of the central problems addressed at a symposium held in 1983 by the Army Institute for the Behavioral and Social Sciences at Texas Tech University in Lubbock, Texas.

In order to avoid "cognitive freezing" in battle, the academic and military experts at this symposium proposed several different approaches to training, including: "overtraining" so that under stress the desired behavior occurs; creating quantitative instead of qualitative tasks (spraying automatic weapons fire instead of aiming single shots at people); and forming strong peer groups. Also, the use of hypnotism and drugs has been hinted at (Hunt and Blair, 1985). Officers, technical specialists, and even the average soldiers to some extent will have to be helped to be innovative and show initiative, two qualities high-tech weapons, in general, and AirLand Battle (ALB), in particular, depend upon. Many analysts see such conflicting advice as indicative of the unrealistic assumptions of ALB, which is probably true. But it also shows that postmodern war calls for more than one type of soldier. At a minimum there must be those who can perform almost mindlessly under extreme conditions that most humans cannot bear, and at the same time there is a need for experts in management, technical repairs, and the application of weapon systems.

Psychotechnologies: Be All That You Can Be Made Into

> Eyes of men running, falling, screaming
> Eyes of men shouting, sweating, bleeding
> The eyes of the fearful, those of the sad
> The eyes of exhaustion, and those of the mad.
> Eyes of men thinking, hoping, waiting
> Eyes of men loving, cursing, hating
> The eyes of the wounded sodden in red
> The eyes of the dying and those of the dead.
> —*Anonymous*[4]

Since the Spanish–American War more U.S. soldiers have been lost to psychiatric collapse than have been killed in action. Recent wars have seen the rate climb to twice the number killed and almost a third of all casualties. In a conventional war in Europe between Great Powers it is estimated that 50 percent of the casualties will be psychiatric. On a nuclear (or chemical/biological) battlefield they will certainly be extremely high among survivors. It is the escalating speed and lethality of battle that is seen as the cause of these increased psychic casualties.

To deal with this problem, the U.S. Army set up the Human Resources Office (HUMRRO) and other think tanks. HUMRRO coined the term "psychotechnology." It has sought to apply its vision of "human engineering," "human quality control" and the "man–weapon system" to U.S. military problems since its founding in 1951. But this systems view is just the beginning. The human component must be modified if it is not to be the weakest link in an integrated weapons system. The incredible demands of postmodern war have precipitated a bureaucratic scramble for technological solutions.

Among the most significant projects for understanding the role of the future soldier are those to introduce artificial intelligence (AI) into the cockpits of Air Force planes and Army helicopters. Through the Army–Air Force program to build the AI Virtual Cockpit, and DARPA's "phantom flight crew" of five expert systems in the SCP's Pilot's Associate demonstration project, the pilot will be intimately connected to his or her flying–fighting machine.

The goal is to improve "man–machine interaction," including "control through line of sight, voice, and psychomotor responses." Computerized "decision aids" are proposed, along with a vague call for techniques to "effectively and efficiently couple operators to advanced systems."

Kenneth Stein reports that "all on-board systems" will be "monitored and diagnosed" for their "health and current/projected operational status." This includes the human pilot who is prone to blackout, redout, exhaustion, wounding, or distraction. The "tactical planning manager . . . may wait for pilot confirmation or may initiate responses" itself. "Where there is time for the pilot to make a judgment" it will graciously suggest possible actions and recommend one (Stein, 1985, p. 73). The monitoring program has been dubbed "the guardian" (Editors of Time–Life, 1988, p. 81).

This "pilot state monitoring" is slated to include systems that read the pilot's brain waves, follow eye movements, and test the conductivity of sweaty palms. All this is in order to gauge his or her mood so the computer will know how to communicate with the pilot, or even when to take over the plane from the pilot if it is deemed necessary (Faludi, 1986; Wilford, 1986).

Work is also proceeding on a number of subroutines that will plan the Pilot Associate's possible future duties by predicting the pilot's requests and actions ahead of time. As a team from the Artificial Intelligence Laboratory at the Air Force's Institute of Technology coyly notes: "Because the pilot and computer have trained together, the pilot expects the computer to begin problem solving when the situation dictates it." The team goes go on to propose a "goal detector" to "deduce" the goals of the pilot "by observing his actions" (Cross et al., 1986, pp. 152, 163). According to some of the other Virtual AI Cockpit researchers, this "mindware" is supposed to create "a fully interactive virtual computer space" wherein the human and the computer "live together" (Wilford, 1986).

For the more distant future the military is clearly aiming for direct brain–computer connections. It could be through computer monitors "reading" the pilot's specific thoughts. The Navy has been sponsoring such studies since at least 1970. A report on some of this early research is called "Electrical 'Windows' on the Mind: Applications for Neurophysiologically Defined Individual Differences." This survey claimed, "Real progress toward a more fine-grained window on the mind began when digital computers became generally available" (Callaway, 1976, p. 90).

A Stanford Research Institute (now called SRI) project in the early 1970s, funded by the DoD, aimed not only at reading minds but also at having a computer insert ideas and messages into the brains of people. Project director Dr. Lawrence Pinneo reported that his computers were 30–40 percent accurate in "guessing" what a person was thinking (*Counterspy* Staff, 1974, p. 5). Research at Johns Hopkins University has managed to sort out enough brain waves to predict when a monkey will move its arms by reading its electrical mental patterns before it acts. At Wright–Patterson Air Force Base they've gone even further. Human subjects have been trained, using biofeedback signals of flickering lights, to fire brain waves in increasing or decreasing amplitudes. These signals turn their flight simulator left or right, as if they were flying by brain waves (Tumey, 1990, p. 16).

Then again, perhaps the connections will also be hard wired neuron to silicon. The Air Force has paid for much of this biocybernetic research since the 1970s, including work in its own labs that puts computer chips into dog brains in experiments aimed at giving pilots "an extra sensing organ." But it is the Army that has paid for the work at West Texas State (and maybe the similar work at Stanford) that has succeeded in growing rat and monkey neurons to silicon chips (Goben, 1987, p. 13; *Scientific American* Staff, 1987, p. 67). The goal is to develop biochips that can be activated by hormones and neural electrical stimulation and which can, in turn, initiate hormonal and mental behavior in humans. It is hoped that such human–machine integration will result in quicker reaction times, better communication, improved control, and greater reliability overall.

Jeffrey Moore, a scientist in the elite advanced weapons group at the Los Alamos National Laboratory, wants to use the biocybernetic brain-computer connection work of the Pilot's Associate and related programs to allow a foot soldier to control a 200-pound suit of armor called PITMAN. It would be capable of stopping a 50-caliber bullet and offer CBN protection. A series of small electric motors would move the massive limbs and a brain-accessing computer would control them. Los Alamos even calls it "a mind-reading protective suit" (Davies, 1987, p. 78).

Even farther out is the research by the Delta Force, who coined the Army's motto "Be All You Can Be." This Delta Force got its name from the belief that technology was the difference, or delta, between the United States and the USSR. (It is unrelated to the antiterrorist commando teams of the same name.) A Delta Force spin-off proposed the First Earth Battalion plan in 1981 for the "warrior-monk" who had mastered "ESP, leaving his body at will, levitation, psychic healing and walking through walls."

Lt. Col. Jim Channon, U.S. Army, the originator of the First Earth Battalion, explains that the United States lost in Vietnam because "we relied on smart bombs instead of smart soldiers." He also states that "stronger than firepower is the force of will, stronger still is spirit, and love is the strongest force of all." But actually Channon's ideal is more in the line of the cyborg. He thinks the future of war belongs to psychoelectronic weapons and explains why the "free world" will have an advantage in developing them:

> If you look for clear examples of where the free world has an advantage over the world of nonbelievers, you will discover two resources that clearly stand out in our favor. They are God and microelectronics. The beauty in that is you can use the microelectronics to project the spirit . . . brains work like that. Hence the field of psychoelectronic weaponry. (quoted in McRae, 1984, p. 124)

Can one ask for clearer proof that at least some military leaders not only have a religious faith in computers but also invest them with deep mystical powers? The urge to create life, to defeat death by dealing it out, and to master nature are so powerful, at least in this case, that military policy becomes stranger than most science fiction.

A number of Channon's ideas on training were investigated by the Pentagon, and the First Earth Battalion had over 800 officers and bureaucrats on its mailing list, including eight generals and an undersecretary of defense. But it was not a major research effort. It got less than $10 million a year (p. 6). More important, it shows how desperately the U.S. military is searching for solutions to the paradoxes of postmodern war, and how even the most 'spiritual' formulations can suddenly be twisted around to become part of the general hypercomputerization of the U.S. military.

This technoscience turn is a common one when the military describes its attempts to find strategic advantage in occult practices. Studying Soviet ESP becomes research into "novel biological information transfer" for the CIA, and psychically tracking submarines is, according to the U.S. Navy, investigating "the ability of certain individuals to perceive remote faint electromagnetic stimuli at a noncognitive level of awareness" (p. 5). The Army has even given out a number of contracts to buy psychic shields for missile silos to prevent psychics from detonating the warheads before they can be launched. Other studies have explored performing and preventing psychic computer programming. As Ron McRae notes in his book on military psychic research, *Mind Wars*, the military point of view is that the "psychic control of computers would indeed be analogous to a nuclear monopoly" (1984, p. 54).

Related projects include SRI's numerous studies, by researcher W. F. Hegge, aimed at "controlling automatic responses to stress and injury" for the "non-drug management of wound-related pain" (Manzione, 1986, pp. 36–38). Another, is the U.S. Army's use, in 1981, of remote-viewing psychics to look for one of its officers, Brig. Gen. James Dozier, kidnapped in Italy by the Red Brigades. Three years later the U.S. military launched *Project Jedi* (yes, it was named for the *Star Wars* movie knights), to see if neurolinguistic programing could improve soldier performance (Squires, 1988, p. 3). All of these programs involve reconceptualizing New Age and occult knowledges with metaphors of information transfer, networking, and programming from computer science. They also share a very low success rate.

In light of the failure of these esoteric methods and considering their social context, it is unsurprising that the U.S. military would put the most energy into developing more direct and traditional (at least in the United States) ways of controlling stress: drugs.

Just Say Yes to Drugs

A U.S. Army (1982) study of future war, *AirLand Battle 2000*, warns starkly: "Battle intensity requires: stress reduction." New antifatigue and antistress medicines that work "without degradation to performance" are a high priority. The noted military psychiatrist Dr. Richard Gabriel claims that this research for "a nondepleting neurotrope" ("a chemical compound that will prevent or reduce anxiety while allowing the soldier to retain his normal levels of acute mental awareness") is nearing success with grave implications:

> Both the U.S. military and the Soviets have initiated programs in the last five years to develop such a drug; the details of the American program remain classified. . . . The U.S. military has already developed at least three prototypes that show great "promise." One of these drugs may be a variant of busbirone. If the search is successful, and it almost inevitably will be, the relationship between soldiers and the battle environment will be transformed forever. . . . if they succeed . . . they will have banished the fear of death and with it will go man's humanity and his soul. (1987, pp. 143–144)

While unique in terms of their potential effectiveness, these current projects to develop such drugs are part of a U.S. research program going back to the 1950s and military traditions thousands of years older.

There is indeed a long history of using drugs to improve performance in war. Dr. Gabriel notes that the Koyak and Wiros tribes of Eurasia used a drug made from the mushroom *Amanita muscaria* that was probably quite effective. He also discusses the *hashshashin* ("assassins"), who used hashish, Inca warriors chewing their coca leaf, and the traditional double jigger of rum for British soldiers and sailors about to face battle. The Soviets combined strict discipline (over 250 general officers shot for cowardice in World War II and thousands of those in the lower ranks executed) with issues of vodka and valerian to calm Red Army soldiers, and caffeine to wake them (pp. 136–140).

Official U.S. interest in drugs other than alcohol can be traced to the 1950s. As early as 1954 the Air Force was testing the performance effects of dextroamphetamine, caffeine, and depressants on men at Randolph Air Force Base, Texas. Three years later the amphetamine was tested again (Payne and Hauty, 1954, pp. 267–273). There have probably been hundreds of studies since.

When it was revealed that the CIA had secretly dosed unsuspecting subjects with LSD and sought ways to psychoprogram killers (with drugs and by using "psychology, psychiatry, sociology, and anthropology"), a giant scandal ensued (Congress of the United States, 1975, p. 610). An important side effect of the resulting investigation was the public unveiling of most military human-subject research up to the mid-1970s. Bearing in mind that each human-subject study was probably based on many more animal studies, it becomes clear that millions of dollars was spent between 1950 and 1975 on the search for drugs that would lower stress and fear while raising or maintaining performance levels.

For example, between 1956 and 1969 the U.S. Air Force sponsored over $1 million in research by doctors at Duke, Baylor, and the University of Minnesota on the effects of stress. This included Dr. Neil R. Burch's work on "Psychophysiological Correlates of Human Information Processing," which he claimed had "direct potential applications to problems of interrogation . . . and may be employed in programing an on-line automatic analysis of data." In other words, he hoped his work would help develop a machine to interrogate prisoners. The guiding principles of this work were clearly stated by the Duke doctors: "to develop specific means to alter of modify responses [of humans to stress] in any desired fashion" (Congress of the United States, 1975, pp. 1135–1138).

In 1971 two researchers at the Army's Aberdeen Research and Development Center and Proving Ground published an annotated bibliography on behavior modification through drugs for the Human Engineering Laboratories, showing the ongoing military interest (Hudgens and Holloway, 1971).

The congressional investigation of 1975, chaired by Sen. Edward M. Kennedy (Dem.-Mass.), revealed that at that time there were at least 20 human-subject studies on controlling stress being conducted by the military. At Walter Reed Army Hospital, there were experiments on the effects of stress on "higher order . . . human performance," "psychophysiological changes . . . of problem solving," and "military performance." At the Army Research Institute of Environmental Medicine, researchers examined "the effects of emotional stress" and assessed "visual response times" under stress. "Metabolic, physiological, and psychological effects of altitude" were measured at the Letterman Army Institute of Research. The Army Medical R&D Command sponsored in-house research on recovery from fatigue. The Navy in Oakland, CA, took a much mellower line and looked into hypnosis for pain relief and relaxation through meditation. While their fellows at the Navy Research Lab did large studies on "effects of combined stresses on Naval Aircrew Performance" and the "psychological effects of tolerance to heat stresses" along with a dozen studies on the effects of cold, water pressure, acceleration, and biorhythms on military performance (Congress of the United States, 1975, pp. 620–627, 640).

The biggest program reported was run by the Edgewood Arsenal and involved an attempt to make a self-administered antidote for soldiers exposed to chemical weapons, specifically oximes. The antidote tested was a combination of atropine, the mysterious TMB4, and the antiphobic drug benactyzine (pp. 765–791).

Considering all this research it is no surprise that the U.S. military issued illegal drugs during Vietnam to elite units and probably still does. As Elton Manzione, a former LRRP (member of a Long-Range Reconnaissance Platoon), reports, "We had the best amphetamines available and they were supplied by the U.S. government" (1986). He also quotes a Navy commando:

> When I was a SEAL team member in Vietnam, the drugs were routinely consumed. They gave you a sense of bravado as well as keeping you awake. Every sight and sound was heightened. You were wired into it all and at times you felt really invulnerable. (p. 36)

Dr. Gabriel argues that major advances in the esoteric disciplines of molecular biology, biocybernetics, neurobiochemistry, psychopharmacology, and related fields means that much more effective drugs can be developed. As noted above, he fears that both the United States and the Russians are about to develop such a drug with the effect of not only keeping soldiers from feeling fear but almost anything else as well, making them functional psychopaths.

Since the postmodern battlefield also requires humans to fight 24 hours a day, there has to be research like Project Endure of HUMRRO to develop night goggles and scopes, or even to modify the human eye itself for night vision, by using atropine (a belladonna derivative) and benactyzine for dilating pupils. The 1978 investigations by Optical Sciences Group of San Rafael, Cal., A. Jampolsky, chief investigator, brought in $700,000 for one of the experiments (Manzione, 1986, pp. 36–38). Of course, infrared and other night vision devices have proven very effective in this regard as well.

The programming of soldiers with drugs is analogous to programming computers. Imagine your immune system programmed against VD (veneral disease), viruses, bacteria, and various toxins; a body with attached bionic parts and eye inserts; a brain hard-wired to a mechanical associate; your mind drugged or psychoprogrammed against stress, fear, altitude, depths, heat, cold, and fatigue; continuously connected and monitored by the computer systems that you watch and use; riding in some secure CBN microenvironment protected by autonomous and slaved weapon systems and controlling vast resources in destructive power and information manipulation: you are a cyborg soldier.

Postmodern Soldiers

> There's a difference between a soldier and a warrior.
> —*Delta Force operator*[5]

> Advances in networking technologies now make it possible to think of people, as well as databases and processors, as resources on a network.
> —*RAND analysts, Arquilla and Ronfeldt (1993)*

> That there will be future warriors is the only certainty.
> —Col. Frederick Timmerman, Jr. (1987, p. 55)

Col. Frederick Timmerman, Jr., U.S. Army, director of the Center for Army Leadership and former editor-in-chief of *Military Review*, embraces this creature with pleasure, naming it the future warrior:

> In a physiological sense, when needed, soldiers may actually appear to be three miles tall and twenty miles wide. Of course in a true physical sense nothing will have changed. Rather, by transforming the way technology is applied, by looking at the problem from a biological perspective—focusing on transforming and extending the soldier's physiological capabilities . . . [can we not achieve the superman solution]? (1987, p. 54)

Consider the range of Colonel Timmerman's speculations. (And they are not just his, actually, but as we have seen they are military policy. They are breathing people with healthy budgets. They are embodied in real weapons and real soldiers.) They don't seem very coherent. They are a search, a recognition, that if war is to remain central to human culture, as it is now, and technological development continues, then soldiers will have to change, even "transform," their bodies and also their role. Colonel Timmerman again:

> It sounds radical, but the time when soldiers are merely soldiers may be ending. Because of an enhanced social role, soldiers of the future may have to be social engineers, appreciate the political implications of their every move and be able to transform themselves to perform missions other than those currently classified as purely military. . . . Finally, we may actually be able to use enhanced social capabilities to degrade an opponent's social cohesion. (p. 53)

The good colonel seems to have slipped through an elision in time. This is what covert soldiers (spies, spooks, mercs, operatives, special forces, commandos, delta force, spetznek) from Central America to Central Africa to Central Asia do already. Today. And didn't NATO's arms race ("enhanced social capabilities") have some effects on the Warsaw Pacts' "social cohesion"? "Language skills" are already a primary weapon through black and white propaganda and public relations.

So in many strange ways certain themes of information, high technology, computerization, and speed remain consistent. When they collide with traditional military culture (as institutionalized in the armies, navies, and air forces of the most powerful world empire ever) these themes produce the many cyborg images and realities of the U.S. military today. But they represent just some of the potential developments of the postmodern soldier.

Even as they come to pass they are inevitably confounded and contrasted with other fragments of the warrior icon, because the same conditions that have forced the U.S. military to reconceptualize itself (especially the advances of technoscience) have led others, soldier and civilian, to appropriate the warrior mythos within an argument that the organized killing of war must end.

This is what the final chapter is about. But before we go there the imagining of future wars, in military planning, must first be confronted. Sadly, all our dreams of peace come from the waking nightmares of war.

Chapter Eleven

Future War: U.S. Military Plans for the Millennium

> We will make our plans to suit our weapons, rather than our weapons to suit our plans.
> —*Robert P. Patterson, Undersecretary of War, October 1944*
> *(quoted in Sherry, 1977, p. 150)*

Futurology and the U.S. Military

Planning for the future has always been a crucial element in organized warfare. The title of the very first chapter of Sun Tzu's *The Art of War* can be translated as "Plans," or "Estimates," or "Reckoning," or even "Calculations" (Griffith, 1962, p. 63). It ends with a discussion of the importance of calculating the probable outcome of any war or battle. Sun Tzu even goes so far as to say, "with many calculations, one can win; with few one cannot." The Chinese character used for "calculations" represents "some sort of counting . . . device, possibly a primitive abacus," according to the translator, Col. Samuel B. Griffith. This shows just how far back the urge to quantify war goes. Griffith thinks that at least two separate logistical calculations were made, one national and one strategic (p. 71).

But with postmodern war the concern for planning the future has, in many ways, become more important than fighting wars in the present. This is certainly true of imaginary and unthinkable war involving superweapons, a war that by necessity consists of only planning and never any fighting, until . . . But even midintensity and low-intensity conflicts are planned, and gamed, and simulated to a degree that would surprise Gen. John J. Pershing as much as Sun Tzu.

This is because postmodern war is predicated on constant technological change involving the continual evolution of new doctrines. As both friendly and enemy forces are changing their weapons, strategy, and tactics all the time, it is necessary to institutionalize the process of worrying about future battles. Gen. Gordon R. Sullivan, Army Chief of Staff, has cleverly suggested that each doctrinal iteration be given software numbers—so that Force XXI, for example, would be 11.0; interim versions would be 11.1, 11.2, and so on until a new release entered the market (1995, p. 14). Planning innovation has to be institutionalized and systematized to accommodate the military bureaucracy.

Modern systematic planning seems to have started with the German General Staff in the ninteenth century. Soon, all the industrial powers had mobilization and strategic plans for potential wars. But they didn't look very far ahead. Officially, there was no planning for new types of wars until the middle of World War II.

Michael Sherry (1977), in his book *Preparing for the Next War*, on U.S. future war plans made during World War II, argues that an ideology of preparedness came to dominate U.S. military planning. At it heart was a certain "technological imperative." As Sherry explains, the position of the preparedness spokesmen was that war now "moved too swiftly to permit scientific research to wait until after the first shot was fired." Because of the rapid growth of technoscience, "invention was shrinking space and collapsing time so abruptly that it imperiled conventional notions of preparedness." Permanent mobilization seemed the only answer (p. 130).

For some, preparedness became a crusade. Consider the position of Edward Bowles, scientific adviser to Secretary of War Henry L. Stimson, on uniting science, industry, the universities, and the military:

> We must not wait for the exigencies of war to drive us to forge these elements into some sort of machine. . . . [Their integration] must transcend being merely doctrine; it must become a state of mind so firmly imbedded in our souls as to become an invincible philosophy. (quoted in Sherry, 1977, p. 133)

The long-range planning during World War II was also significant in that the planners "ignored tradition" and refused "to name a probable adversary." Instead, they argued one of the core ideas of preparedness ideology—that "the primary danger to American security arose from fundamental changes in the conduct of war and international relations, not from the transient threat posed by a particular nation" (p. 159).

Sherry points out that more than mere ideology motivated the preparedness movement:

> Preparedness would help nourish the scientific infrastructure needed for economic expansion, strengthen ties between the corporate and political elites, and defend access to the markets and materials deemed necessary for continued economic growth. And preparedness could stimulate gross

investment at a time when other forms of government spending were anathema to powerful interests. (p. 236)

Knud Larsen (1986) calls this kind of practically grounded moral absolutism a "ritualized ideology." He stresses the social and psychological factors behind the arms race specifically and military technology generally. Larsen's analysis demonstrates that, like much of postmodern war, preparedness ideology is overdetermined. It fills many functions institutionally, personally, and culturally.

Unsurprisingly, it was the U.S. Army Air Force that led the way in institutionalizing preparedness ideology and in imagining future conflicts, and the issue was inevitably framed in terms of technoscience and war. At the tail end of World War II, Gen. Hap Arnold, father of the U.S. Air Force, told his scientific adviser, Dr. Theodor von Kármán, that the alliance between the military and science was both political and strategic. Michael Sherry describes Arnold's analysis and quotes from Arnold's communications with von Kármán:

> The strategic danger to the United States lay largely in the unfolding technological revolution: the United States would face enemies and the possibility of "global war" waged by offensive weapons of great sophistication. But the American response would be shaped by considerations of domestic politics. The United States had to reverse "the mistakes of unpreparedness" prior to World War II, "particularly the failure to harness civilian science to military needs." And technological development would respond to "a fundamental principle of democracy that personnel casualties are distasteful. We will continue to fight mechanical rather than manpower wars." (pp. 186–187)

Starting in 1945 with *Toward New Horizons*, the U.S. military initiated a number of studies aimed at understanding future war in terms of future technologies. To do so they have brought together futurists, scientists, science fiction writers, military officers, and civilian technobureaucrats in a series of conferences. The U.S. Air Force's *Toward New Horizons* was followed with the *Woods Hole Summer Studies* of 1957 and 1958 and then *Project Forecast* in 1963. That was followed in turn by *New Horizons II* in 1975 and by *Forecast II* in 1985 and then by *Innovation Task Force 2025* (Gorn, 1988, p. v).

The other services have had almost as many studies. The approaching millennium has also encouraged a number of new speculations about war in the next century, including *AirLand Battle 2000, Army 21, Air Force 2000, Marine Corps 2000, Navy 21,* and *Focus 21* (a joint Army and Air Force product). Studies of the growing importance of Space War culminated in the formation of a Unified Space Command. Together with reports from elite ad hoc groups, such as President Reagan's Commission on Integrated Long-

Term Strategy, these constitute the future war policy of the United States. Some of the details are quite startling.

AirLand Battle 2000: The Twenty-First-Century Army

The twenty-first century is less than one procurement cycle away, as Capt. Ralph Peters, U.S. Army has written in "The Army of the Future" (1987, p. 36). Most of today's junior officers will serve more of their careers in the twenty-first century than this one, noted a retired general back in 1980 (Sarkesian, 1980, p. vi).

Produced in 1982, AirLand Battle 2000[1] was a "jointly developed concept agreed to by the [U.S.] Army and the German Army." It has since been distilled into the official U.S. Army plans for war in Europe and also recycled as *Army 21*. Even as the possibility of general war in Europe grows more remote, AirLand Battle (ALB) is still important because it represents the model for any conflict the U.S. might fight with a well-armed enemy. It was the strategic doctrine used for Desert Storm.

ALB assumes that future battles will be three dimensional, requiring intimate coordination between land and air–space forces. ALB doctrine is based on maneuver and aggression in the context of a hyperlethal chaotic "battlefield" (battlespace?) of hundreds of cubic miles. NATO plans were for deep counterattacks in the rear of the attacking Warsaw Pact armies immediately after, or perhaps immediately before, an invasion of Europe. Preemptive strikes are an integral part of ALB plans, which rely heavily on air and space power. Despite the miniscule chances of a Soviet attack on NATO, the ALB supplied the pretext for a whole set of new doctrines (of multiarm coordination) and a new suite of weapons (especially remote-controlled and autonomous drones, computer networks, and sensors) that have already been used in Grenada, Nicaragua, Panama, and other LICs (M. Miller, 1988, pp. 18–21), as well as the midintensity war in Kuwait and Iraq.

AirLand Battle 2000 puts forward technological solutions to the problems expected to result from the mind-rending impact of high-tech weapons in continuous combat lasting days at a time. Many units will be obliterated; others, merely broken. All evidence suggests that most soldiers will not be able to fight with much effectiveness under these conditions (Hunt and Blair, 1985). To deal with this battlespace, *AirLand Battle 2000* advocates improved support services (medical, logistical, and even such marginal aids as video chaplains, talking expert systems to give legal advice, and computer war games for "recreation and stress reduction") and stronger measures that will help to integrate the individual soldiers into parts of a complex fighting unit.

See-through eye armor and other "personal bionic attachments to improve human capabilities" are planned to go along with artificial bones, artificial blood, and spray-on skin for the wounded. For those with major wounds, WHIMPER (Wound-Healing Injection Mandating Partial Early Recovery) shots would allow their evacuation or even their quick return to combat. Universal antiviral, antibacterial, and anti-VD vaccines are to be developed along with mycotoxin antidotes. A universal insoluble insect repellent will supposedly save time and reduce aggravation, as will chemicals to stunt hair growth, retard body functions, and keep teeth clean without brushing for six months at a time.

Miniature sensors will warn soldiers of chemical, biological, or radiological threats. Miniature data discs will hold each soldier's records. Other "automated devices" will judge his or her "physical and psychological fitness" and decide who must keep fighting.

The Strategic Computing Program (SCP) worked on an expert system battlemanager to advise ALB commanders on the corp level. It was to predict enemy activity, track and filter the extraordinary amount of information that battle now generates, advise the humans, and even issue their orders. It is expected that satellites in near space will direct individual artillery rounds, send messages between commands, and pinpoint every single friendly soldier and machine. Such is the Army's dream for managing postmodern war: total information about every logistical or fighting machine, human, and system.

The U.S. Air Force has gone further than the Army in differentiating various roles for machines, humans, and cyborgs: machines are for the mindless brute work; humans are still needed for some of the management, maintenance, and manipulation of weapon systems; cyborgs (human–machine weapons systems) are for fighting. In the future the Air Force plans to take this approach to the technoscientific limits, as can be seen in its own blueprint for the next century.

Project Forecast II: The Twenty-First-Century Air Force

> Problems never have final or universal solutions, and only a constant inquisitive attitude toward science and a ceaseless and swift adaptation to new developments can maintain the security of this nation.
> —Theodor von Kármán, 1945 (quoted in Gorn, 1988, p. 37)

Michael Gorn has written a history for the U.S. Air Force of its science and technology forecasting (1988). In it he discerns a number of important trends. The most significant is that over the course of the major Air Force futurology projects the role of independent civilians has continually declined, while the importance of military personnel with science degrees and of scientists working directly for the DoD has increased proportionately.

Of all the services the Air Force has had the greatest interest in predicting the technological future because it sees air power as being integrally linked to science. So it was out of the first futurology study, *Toward New Horizons*, that the USAF Scientific Advisory Board came. The same man who led *Toward New Horizons*, Dr. Theodor von Kármán, also chaired the first Advisory Board and directed the *Woods Hole Summer Studies*. His first futurological analysis for the Air Force was a wartime study of German and Japanese technologies, *Where We Stand*, which included certain recommendations for future research. It was published as one of the 12 *New Horizon* volumes. Volume 1 of the series was entitled, *Science, the Key to Air Supremacy* (pp. 30–50). Gorn points out that, despite von Kármán's desire to keep independent outside scientists in control of predictions and scientific advice, there has been a continual erosion in both these areas. Control of futurology studies has been shifted from the Scientific Advisory Board to various Air Force bureaucracies, such as the Air Research and Development Command and the Air Force Systems Command. Meanwhile, the Scientific Advisory Board has shrunk and lost most of its civilian participants, while its mandate has become the analysis of specific immediate technical problems on an ad hoc basis.

This militarization of Air Force official futurology has had several effects including a weakening of technical skepticism, since all projections come not from practicing scientists but from active weapons engineers and developers. As Gorn notes, "One element was lacking in enlisting Air Force officers for long-range R&D reports: True disinterestedness toward the subject matter." Consideration of "the relationship between proposed technologies and their place in the general defense landscape" has also disappeared (pp. 185–186). Finally, Gorn quotes Gen. Hap Arnold, who told von Kármán that only independent scientists could solve the military's most difficult technical problems: "the technical genius which could find answers . . . was not cooped up in military or civilian bureaucracy but was to be found in universities and in the people at large" (p. 268).

The Air Force has led the way in institutionalizing postmodern war, especially the role of science and the innovation of innovation. In their very first study of technology and future war, von Kármán and his associates successfully advocated a number of reforms to establish science throughout the Air Force by:

- Setting up a Scientific Advisory Board and science offices in commands such as intelligence and headquarters
- Funding a large R&D program with connections to university and industrial laboratories
- Founding new Air Force research labs
- Training significant numbers of officers in technical and scientific disciplines

Other proposals that were either immediately accepted or eventually implemented included the development of electronically assisted and purely automated weapons (bombs, missiles) and platforms (planes) (pp. 37–40).

Still, much bureaucratic infighting was necessary to keep scientific and technical innovation in a leading role within the Air Force. In 1947, for example, one of von Kármán's aides and friends, Maj. Teddy Walkowicz, had to appeal to von Kármán for help in keeping the Scientific Advisory Board from eclipse. In a letter he wrote to von Kármán he warned, "If the pilots reign supreme in peace time as they do in war time the whole cause will be lost . . . and the . . . tragic course of any future war will be decided long before the first shot is fired" (p. 47). Over time such struggles faded as many institutions within the Air Force and outside it became dependent on continual technoscientific innovation, although independent civilian scientific input has continued to fade, as Gorn's book chronicles.

Later Air Force studies continued with such proposals and also began to focus more and more on sophisticated communication systems for improved command and control, including space systems, and for goodies such as digitalized, worldwide cartography. The most recent full study by the U.S. Air Force (1986), *Project Forecast II*, is the best example of where these projects have been heading.

Project Forecast II was generated by a large collection of experts: 175 civilian and military researchers divided into 18 technology, mission, and analysis panels. Indeed, Gen. Lawrence Skantze, Commander, Air Force Systems Command, makes a convincing case in his briefing at Aerospace '87 that *Project Forecast II* has shaped current Air Force R&D to a great extent (U.S. Air Force, 1987). He claims that in 1987 the project's suggestions absorbed over 10 percent of the $1.6 billion Air Force Laboratories' budget (his command), with a similar level of expenditure being planned up through 1993. The Air Force as a whole kicked in another $150 million for 1988. In 1987, a group of 24 key aerospace companies ponied up $866 million of their own for the *Project Forecast II* proposals, 44 percent of their $2 billion of internal R&D.[2] This "private" research on military proposals doesn't officially count as military work, although it is along military lines, with military specifications, aimed at winning military contracts, and only possible because of the profits from earlier military work.

Easily over $1 billion in fiscal year 1987 was spent on R&D on the proposals from this conference, with $1.2 billion or more slated for 1988. Clearly, "The United States Air Force is committed to implementing the results of *Project Forecast II*," as the executive summary proclaims on p. 1. What are these results—or, more precisely, what's on this wish list?

Here's a sampling. New materials are proposed. The use of photons in place of electrons in computers is advocated to speed them up and make them harder to disrupt electromagnetically. There is the dream of "optical kill mechanisms"—lethal lasers to blind and kill sensors and people on the battleground. Many of the proposals are to help fulfill the Air Force's desire for "rapid, reliable, and affordable access to space." Others are members of "a family of weapons which autonomously acquire, track, and guide to a broad spectrum of air and surface targets in all environments." Specific examples include low-cost drones for surveillance, homing in on radar sites, or carrying their own "smart" antiarmor weapons with a "fire 'n' forget terminal maneuvering" capability. "Smart skins," combining sensors, new materials, and computers in a system capable, they hope, of a "total situational awareness," will be used to cover both independent brilliant weapons and piloted aircraft.

Continual support for the virtual cockpit and the pilot's associate are also strongly advocated. One area of current interest that is somewhat slighted is LIC. The same cannot be said for the other high-tech service and its relatively low-tech junior partner. Both the Navy and the Marines have made future LICs their primary concern.

Navy 21 and *Marines 2000:* Force Projection in the Twenty-First Century

> There won't be any noncombat areas in the world. It will all be a potential combat zone.
> —Navy 21 *study (U.S. Navy, 1988)*

Navies have often been the long-arm of empire. It is a tradition that seems likely to last into the twenty-first century. As the former Chief of Naval Operations put it, with some standard stereotypical assumptions,

> You can tell these people in the Middle East or Africa that there are eight or ten men sitting in a silo in Montana. They don't give a damn about that. They can't see anything. They don't know where Montana is. But if you say, "Look at that big ship, out there," it has an impact. (Adm. C. R. James, quoted in Fraser, 1988, p. 53)

By the late 1980s, many in the U.S. Navy understood that as the Soviet Union weakened, the United States would be playing a more aggressive role in the Third World. "The next naval battle we fight likely will occur in the Persian Gulf, Mediterranean, or Caribbean," one officer accurately predicted in 1988 (Morgan, 1988, p. 58). But despite the low quality of the likely opponents, high-tech weapons such as Tomahawk cruise missiles are crucial because this is "an era when the loss of even one aircraft in an offensive strike may be politically unacceptable" (Fraser, 1988, p. 54).

So nervous is the Navy about the growing importance of LIC that it has converted one of the new *Seawolf* class of fast-attack nuclear submarines (SSNs) into the most expensive covert operations vehicles ever. The main mission of attack subs has been to destroy Soviet submarines, but with the decline of that threat the need for SSNs has rapidly diminished. In a 1988 article, Lt. Cmdr. Marcus Urioste advocated using SSNs for landing special forces, land attack with cruise missiles, direct surveillance of harbors and coastlines, and delivering robotic vehicles and remotely piloted vehicles and sensors for covert intelligence gathering. He even argued that they are a powerful "psychological" threat in midintensity conflicts as well. The commander left the Navy to work for General Electric's submarine program. By 1995 many of his recommendations had been carried out. Obviously, the real danger to him is a cutback of strategic submarines, so other missions must be found. Otherwise, as he admits, they "may be pressed to justify the modern SSNs' costs for use solely in a superpower confrontation" (Urioste, 1988, pp. 109–112).

The U.S. Marines, obviously, have less to worry about from a shift to low- and midintensity conflicts than does the nuclear submarine force. While the Marines have traditionally been the least high-tech of the armed services, lately they too have been seduced by the promises of microelectronics and other fruits of technoscience. Their over-the-horizon amphibious strategy depends on hovercraft and tilt-rotor aircraft—and, of course, accurate battlefield intelligence. Col. Lawrence Karch, in an article on "The Corps in 2001," is gushing by the end of his piece that, "advanced microelectronics" are "the key technology of precision weapons" which will have "science fiction-like capabilities." He concludes by calling for increased innovation in the Marine Corp. It "must," he proclaims, "continue to evolve and not ossify in its thinking." It also must dispense "with the obsolete," retain "the useful," and "acquire the necessary," while "knowing the differences." Clearly, the key component of war, even for the Marines, is information (Karch, 1988, pp. 40–44).

Cmdr. Thomas Keithly, U.S. Navy, sees a central role for information and information processing. He cites "experts" who "expect at least a 1,000 percent increase in the rate of information exchange." He himself predicts future ships will have rapid computers, fiber-optic networks, optical processors, photonics, and neural networks "to handle data much faster by means of parallel and adaptive processing by using light in place of electrons." And for the "basic problem" of how to handle the data, there will be "data integration, and tactical decision aids" that use artificial intelligence (AI). He even claims that improvements in "command and control will result ultimately in mastery of the radio-electronic spectrum" (Keithly, 1988, pp. 52–54).

The new Aegis destroyers, the *Arleigh Burke* class, are highly computerized. Not only do they have the Aegis system and computer-controlled engines, but these systems, are linked by a five-path data-multiplex system that ties together the six stand-alone microcomputers that make up the control system, which in turn "monitors and controls the status, health, and commands to just about everything below the main deck" (Preisel, 1988, pp. 121–123). The *Arleigh Burke* class destroyers also are the first fully pressurized U.S. Navy surface combatants. The hope is that this will help keep nuclear and chemical contaminants out (Morgan, 1988).

As Cmdr. J. Preisel Jr., describes it,

> Everyone in the ship is talking in terms of bits, bytes, and multiplexing systems. Digital logic, with its "and/or" gates, is discussed over chow. This is the engineering department of a naval warship, but it sounds like we have slipped on board the Starship *Enterprise*. Computers have arrived full force in the engineering departments of today's surface navy. (1988, p. 123)

It also sounds a lot like today's Air Force. The Navy is pursuing the idea of stealth ships with a low profile and radar-absorbing coverings. The number of crew needed per ship has been falling for 20 years, and it will fall still further, maybe even to zero a few optimists claim. Drone and semiautomated ships would make it possible for the Navy to have 1,200 ships instead of the 600 planned for the twenty-first century (Keen, 1988, p. 97). Realistically, while there will probably be an increase in the use of air and sea drones, autonomous weapons on ships, AI data sifters, intelligent mines, and so on, crewless ships are hardly likely. But drastic changes do seem in store. One Admiral has proclaimed, "I want the bridge of the next surface combatant—if it has a bridge!—to resemble the cockpit of a 747 aircraft . . . with room for no more than one or two people!" (quoted in Truver, 1988, p. 72) There is even a study group for a paperless ship that would do away with all the manuals and memos that today make up the lifeblood of the Navy. This whole process of reconceptualizing the Navy in terms of new technologies and the old imperial mission is a tremendously complicated bureaucratic dance, which is described in excruciating detail in several articles (Keithly, 1988; Nyquist, 1988; Truver, 1988). But again, what it all comes down to is information.

The *Navy 21* study, done by the Naval Studies Board in 1988, argued quite directly that future war is "information war." One crucial key to winning information war is space, especially space-based surveillance, navigation, communications, and weather satellites. It concluded that the "Navy must control space if it hopes to command the sea." This led to advocating putting antisatellite weapons (asats) and small satellites (lightsats) on submarines and surface ships as well as giving all ships direct satellite downlink capabilities. One of the report's major conclusions is that space is as important to the

Navy as ship modernization (*Military Space* Staff, 1989a, p. 3). But it isn't just the Navy that is enamored with space. For all of the armed services, and the spy agencies as well, space is the place to achieve military superiority.

Space Command

> It is possible that the first—and perhaps the only—battle waged in the next war will be one of information, and this battle may unfold over access to space.
> —*Vice Adm. Jean Chabuab, French Navy*

> I'm unabashedly optimistic about the future of military space.
> —*Gen. Thomas Moorman, Air Force Space Command*[3]

A military role in space is not just some future plan. NASA's space program is incredibly militarized. For example, in the ten years between 1978 and 1988, of the 99 astronauts 64 were active-duty military officers and 9 were retired officers. The active-duty personnel return to their parent service after seven years or two to three flights. The space shuttle has been committed to numerous military missions. Twelve of the 50 shuttle flights taken by 1991 were totally military and all have some military missions. Besides the many military payload specialists who go up to launch milsats and do military experiments, the services have some special plans as well. The Air Force wants to send up a weather expert. The Army wants to send a geologist and a battlefield observer. The Navy plans for oceanographers and communications specialists. All of NASA's projects have been helpful to the military to some extent, but the rush for military space has made the shuttle program the most militarized NASA project by far (Cassott, 1988, pp. 6–8).

Official military space expenditures already take up 5 percent of the DoD budget, and that percentage is expected to double in the next ten years. The goal of this expenditure is "space control." While there hasn't been any fighting in space just yet, space has been part of terrestrial warfare for a number of years now, not just in terms of the satellites and ICBMs of nuclear war but also because of the role of satellites for reconnaissance and communication in smaller scale conflicts. The DoD has claimed that satellites "leverage" U.S. forces by a factor of 4 or 5. This means, in plain English, that they claim that satellites make U.S. soldiers four or five times more effective. This certainly remains to be proven, and the evidence from Grenada and Panama is that it just isn't true (*Military Space* Staff, 1990b, p. 1).

Only with the Panama invasion did space become integral to actual U.S. military operations. Even more space systems were used in the Gulf War of 1991. All in all more than 50 satellites played a part in Desert Storm. The visual satellites, code-named KH-11 but also known as Keyhole spacecraft, are able to see a glove in the desert but can't count the fingers. A radar-imaging satellite called Lacrosse took "pictures" at night and through cloud cover. Other satellites scanned Iraq with infrared, eavesdropped on electronic communications (Magnum and Vortex satellites), watched the oceans (Parcae satellites), and supplied the UN allies with communications channels and positioning information (Frederick, 1991, pp. 2–3).

This war showed how the military space establishment has grown into quite a large creature. At its head is the Defense Space Council, which gives military input to the National Space Council. The top civilian administrator of military space is the Air Force's assistant secretary for space systems, who also heads the National Reconnaissance Office, which directs classified space intelligence programs run with the CIA and NSA. Next in the chain of command comes Unified Space Command and then the various services' space commands. Also there are the Naval Warfare Command, the Air Force's Space Systems Command, which does R&D and the Army's Strategic Defense Command, which under SDI is developing ground-based space weapons. Space research and operational centers have proliferated.

The Air Force's Space Command, for example, directly controls the Cheyenne Mountain Early Warning Center, Patrick AFB, Cape Canaveral, and Onizuka AFB as well as many of the launch sites at Vandenberg AFB. The Air Force's Space Systems Division controls parts of these institutions and the Onizuka Consolidated Space Test Center and the rest of Vandenberg AFB.

Along with all the military satellites and research centers (which includes DARPA facilities and labs of the other services' research arms), the military has a large stake in NASA projects, most notably the shuttle as mentioned above, the Space Station, and perhaps someday the Space Exploration Initiative (SEI) of the Bush administration, which aims to get humans back on the Moon by 2000 and to Mars by 2019. What the military hopes to get from this is help in financing and developing its Advanced Launch System and also some vague role in guaranteeing "Freedom of Space" beyond Earth's orbit.

Closer to home the military has begun to seriously worry about the proliferation of missile capability to Third World nations. The CIA has estimated that by the year 2000 six developing countries will have ICBMs and nine others will have intermediate-range missiles. This threat has become a major justification for continuing SDI (*Military Space* Staff, 1990b, p. 3). The Pentagon has set up a Proliferation Counters Group to worry about this, and they admit that the spread of missile technology will seriously affect "our strategy for intervention in [Third World] areas" (*Military Space* Staff, 1990c, p. 3).

Perhaps the Pentagon's most important futurology study was the report of the Commission on Integrated Long-Term Strategy, described in detail below. One of its cochairs, Albert Wohlstetter, a key adviser to then Vice President Dan Quayle, has stressed that a very important conclusion of the commission is that space must be totally militarized. "Space will be no sanctuary," he promises, "but will be an important determinant of the outcome of the war." The Commission strongly came out in favor of asats, lightsats (but not cheapsats), better tacsats, and more fatsats, if they are survivable. ("Sats" is short for satellites; "a" means anti; "tac" means tactical; "light" means small; "fat" means big; "com" means communications; "mil" means military.) It even argued for possibly staking claim to hunks of space in arcs of up to 15 degrees around key space weapons and making them free-fire zones (*Military Space* Staff, 1989b, p. 3).

The commission's Regional Conflict Working Group called for communications and spy satellites to fight the drug wars and even to control unmanned vehicles for eradicating fields of illegal substances. They also advocated an increased role for space technology in all LICs. All these recommendations the commission included in its final report (*Military Space* Staff, 1988d, p. 8).

Space is the high ground. For the Army little more needs to be said, although the idea is taking awhile to sink in. In the Air Force, despite the resistance of the pilots, it is clear that a commitment to "aerospace power" is behind their serious investment in militarized space. Even the Navy is more and more coming around as various analysts argue that space power is connected to sea power. Usually, space power is seen as a way of preserving sea power, but for some the equation is better if reversed. Adu Karema, a civilian scientist working on naval space systems, claims unconvincingly that "Control of the sea will be more important than ever since such control is necessary to guarantee our access to space or deny access to our enemies" (*Military Space* Staff, 1988c, p. 8).

What all this adds up to is lots of satellites. A congressional report in 1988 said that the Pentagon planned to have 150 satellites in orbit by the year 2015, not counting the 10,000 or so needed for SDI. As Space Command is already having trouble controlling its satellites, this seems a very ambitious program (S. Johnson, 1988, pp. B1, B4).

So space is the final war frontier for all the services. But what is the overall plan for twenty-first century war? President Reagan appointed a commission to determine just that. It "integrated" the various Pentagon and intelligence proposals for future war.

Integrated Long-Term Strategy

In one sense U.S. military strategy, at least as far as technoscience is concerned, is a wild grasping at any new possible weapon. Whenever a scientific discovery is made or even predicted, the Pentagon is there seeking ways of turning it into new weapons. In computing, this can be seen with the massive support for neuro- and parallel computing research, both heavily financed by the DoD. Danny Hillis, the inventor of the connection machine, even claims that "The military is paying for the development of all the interesting parallel computers" (Brand, 1987, p. 193).

But this faith in technoscience extends beyond computing to almost every traditional science fiction weapon including force fields and death rays (S. Johnson, 1986, pp. A1, A17). Superconductors, the military hopes, will allow for all sorts of killing rays and spy beams. After the superconducting discoveries in 1987, military spending on superconductivity in the United States jumped immediately from $5 million to $12.5 million, and it was predicted to hit $150 million within a few years (S. Johnson, 1987, pp. A1, A24).

Yet this technophilia is not mindless. Whatever new technologies seem to come out of these spending frenzies are put to work in a much more important arena than actual war; they are deployed in justifying war. It is a sign of the maturity of AI research, and its centrality to military discourse, that it supplies the basic rationale for structuring and justifying U.S. military policy into the twenty-first century.

In 1988, a very important report was made on what the long-term strategy of the United States should be. The cochairs were the recently retired undersecretary of defense, Fred Ikle, and Albert Wohlstetter, the one-time RAND analyst whose military consulting has made a him millionaire. The panel also included a former secretary of state, Henry Kissinger, a former national security adviser, Zbignew Brzezinski, a former chairman of the Joint Chiefs, Gen. John Vessey, a former NATO commander, Gen. Andrew Goodpaster, the hard-line academic Samuel Huntington, and others only slightly less famous.

The main thrust of their report, entitled *Discriminate Deterrence*, is that new weapons must be developed to enable the implementation of a new doctrine of "discriminate" nuclear war. These new weapons would be "smart" conventional and nuclear missiles. Why is such a doctrine needed? The surface argument is that massive deterrence is not credible. Would the United States destroy Europe, maybe even the world, if Russia invaded Norway? So instead of a general nuclear war, the report calls for a conventional and limited nuclear response using thousands of smart weapons. But actually there are other factors besides a fear of a Russian invasion of Norway that shape the reports' recommendations.

First, there is politics. Wohlstetter admitted during an interview that "It became politically possible to say these things, because it was politically necessary" (Stewart, 1988, p. 1). The political necessity came from the growing irrelevance of NATO in the face of the collapse of the Soviet empire. The report deemphasizes NATO to an amazing degree, especially for 1988.

The second important factor is the recognition that the hegemony of the United States and the (former) USSR is weakening. The future will be "a far more complicated environment than the familiar bipolar competition with the Soviet Union," the report warns. It also predicts that in the twenty-first century over 40 countries will be able to build their own advanced weapons, including chemical, biological, atomic, and missile delivery systems. Along with this threat the United States will face "a broader range of challenges in the Third World" that will require highly mobile forces and a downgrading of the European theater. Since the Cold War with the Soviets has ended, it is not surprising that the "wise men" who wrote the report advocate terming LICs a form of "protracted war." When possible, these threats should be met with U.S.-financed, -trained, and -armed proxy forces, supported from afar with long-range precision-guided weapons.

Finally, the third rationale behind this worldwide version of Vietnamization is the claim that "a microelectronic revolution" makes conventional weapons as effective as nuclear weapons in many cases:

> The much greater precision, range and destructiveness of weapons could extend war across a much wider geographic area, make war much more rapid and intense, and require entirely new modes of operation. (from the report *Discriminate Deterrence*, quoted in Weiner, 1987, p. A1)

Basically, the report is a call for the relegitimation of warfare. LICs are to be rehabilitated as "protracted war" and fought with high-tech computer weapons. Even limited nuclear war is to be considered a viable option, officially only in Europe, but obviously with any of the potential 40 or so middle-ranking powers as well. There must be wars, the report seems to say in many different ways, and computers will allow us to have them without destroying the world or domestic support for empire. Info/cyber/netwars are just the latest flourish of this year's revolution in military affairs.

Just how far this thinking can go is shown by the Pentagon's planning for World War IV. Starting in 1981, shortly after Ronald Reagan took office, the DoD began planning on how to fight and win a six-month global war and still be ready to fight another one, mainly by using computers:

> Long after the White House and Pentagon are reduced to rubble and much of civilization is destroyed, the strategy calls for computers to run a war no human mind could control, orchestrating space satellites and nuclear weapons over a global battlefield. (Weiner, 1990)

The plans for World War IV depend on a number of systems: Milstar satellites, command tractor trailers, a nationwide network of 500 radio stations, and a working army of robots "that can gallop like horses and walk like men, carrying out computerized orders as they roam the radioactive battlefront." These robots will take orders but, in DARPA's words, they will "not generate discourse." Almost all of this World War (III and IV) planning is top secret, hidden away in secret budgets. But what can't be hidden is the incredible commitment to war that such plans demonstrate. Even after a nuclear holocaust, in the face of almost all scientific and humanistic thinking, most military professionals think that there still must be war—but not all of them do. Some are beginning to think that there must be an end to war, as we shall see in the next chapter.

Chapter Twelve

Future Peace: The Remaking of Scientists and Soldiers

The Politics of Science

Science is the most prestigious worldview today in both the developed and the undeveloped worlds. Marxist or capitalist, Shiite or Sunni, Protestant or Catholic, all pay a certain homage to science. Many people, especially scientists, give it the role of their personal defining ideology or religion. Therapies, policies, and countless other systems seek its mantle, from astrology to psychiatry to Stalinism to UFOlogy. Whether or not they are indeed sciences depends on one's definition of a science, but they wish to be.

The legitimizing power of science has been claimed by many political viewpoints. This is certainly true of most variations of capitalist ideology. Whether sympathetic to the state or not, from Democratic liberals to free-market libertarians, the vast majority of the ideological supporters of capitalism give science a central role in their cosmology.

But this is just as true of almost all of the Western alternatives to capitalism, such as early anarchism and almost all brands of Marxism. Peter Kropotkin, a world-famous geographer in his own right, argued that science supported anarchism and anarchism was scientific (1970). While Mikhail Bakunin also felt undistorted science supported anarchism, he was less naive about the possible misuse of science's material and moral power. He warned against the authoritarian tendencies of scientists and the misuse of science by governments. Most importantly, he validated the importance of unscientific abstractions, such as passion and justice, for analyzing and changing the social system (1953). Postmodern anarchists are very skeptical of science, and many have fully taken up the critique of the postmodern feminists and anarchist theorists such as Paul Goodman, who even in the 1950s disputed science's claims to normative truth and disinterested good (1964).

On the other hand, most Marxists still argue that Marxism is scientific, although there are exceptions (see Fee, 1986). Marxism-as-science is a claim that has even convinced some scientists. Between the World Wars many prominent British scientists were Marxists, such as J. B. Haldane and J. D. Bernal, and they even went so far as to argue that future science could only be communist. Stalinist science policy, as exemplified by the Lysenko affair, eventually dissuaded them of this particular delusion. But, in general, it seems illusions about science's total applicability led these ex-Stalinist scientists to continue their arguments for the scientific management of society and for the expansion of science into politics in all its forms, especially warmaking.

Ironically, the well-organized antimilitarist groups of scientists, such as the Cambridge Scientists' Anti-War Group and the Association of Scientific Workers (ASW) in Britain, were led by Bernal and others into the heart of the British war effort in the late 1930s. The leadership of the ASW even collected a list of scientists willing to do war work, despite numerous protests from scientists with more consistent antiwar politics. Finally, the ASW came to the contradictory conclusion that while "war [is] the supreme perversion of science" the danger of "anti-democratic movements" which were a threat to "the very existence of science" meant that "we are prepared to organize for defence" (McGucken, 1984, p. 159). For them, a perverted science is better than no science at all.

Many U.S. scientists also campaigned for a greater role in government decision making, especially on military issues. Along with access to policy making, the scientists usually proposed scientist-controlled institutions to hand out government money. They wanted money without accountability, which didn't happen except in limited cases such as the National Science Foundation. Usually scientists did get lots of money, but at the price of becoming accountable to the military.

Is it any wonder that science has survived, indeed triumphed, in today's perverse form? Well-meaning radical scientists played a major role in the creation of the postmodern war system through the application of scientific methods to social and military processes (Hales, 1974). It was under the mistaken impression that the Nazis were close to developing atomic bombs that a number of antimilitarist, even pacifist, scientists conceived and created the first atomic bombs, unleashing the possibility of annihilation upon the world. That the advances of science and the developments of war may have led to our present situation anyway does not lesson the guilt of the physicists, as many freely admitted. Even J. Robert Oppenheimer said, "Physicists have now known sin."

Still, he didn't feel that guilty. Witness his postwar advocacy of tactical nuclear weapons as a counterweight to development of the hydrogen bomb being pushed by Edward Teller and John von Neumann. Oppenheimer's general evasion of responsibility, famous public quotations such as the one above notwithstanding, can be seen in this letter in 1946 to Vannevar Bush defending the general amnesty given to German scientists and their recruitment into the U.S. military-research apparatus (while a similar integration was taking place on the Soviet side of the Iron Curtain with the scientists they had captured):

> You and I both know that it is not primarily men of science who are dangerous, but the policies of Governments which lead to aggression and to war. You and I both know that if the German scientists are treated as enemies of society, the scientists of this country will soon be so regarded. (quoted in Kevles, 1987, p. 338)

This begs the question: sure, atomic bombs don't kill people, people kill people, but do you give atomic bombs to killers?

By the end of World War II many scientists congratulated themselves on science's integration into government. Some, such as the technocrat physicist Arthur Holly Compton, a Nobel laureate, thought that all of culture should be subordinated to science. He argued that "Only those features of society can survive which adapt men to life under the conditions of growing science and technology" (quoted in Sherry, 1977, p. 128).

Scientists also began pushing for a more worldwide policy role for scientists through the World Federation of Scientific Workers, founded in 1946, and UNESCO, whose first director was Julian Huxley, a leading British liberal. Most capitalist and communist scientists were equally enthusiastic about this alliance. Since 1946 there has been an unprecedented number of scientists in public policy positions. What's seldom discussed about these scientist-politicians is that one of the main reasons they have a voice in public affairs is because of their ability to make weapons. But this is not surprising. In many ways war is the dirty secret of science.

Consider how brief the treatment of this relationship is in philosophy. Such famous philosophers of science as Thomas Kuhn, Hilary Putnum, and even the dada-anarchist Paul Feyerabend hardly ever confronted it. In computer science, a discipline mothered and fathered by war and the military, philosophical works like Margaret Boden's ignore this genealogy altogether; which is all the more startling when one considers how open and debated it is among computer scientists themselves. But there are exceptions. David Dickson's *The New Politics of Science* (1984) is an excellent overview of U.S. Cold War science. Not only does he always keep the militarized context of late-twentieth-century U.S. science clear, but he also situates it within an intense capitalistic discourse that turns individual scientists, government agencies, and nonprofit universities into profit-seeking organizations for all intents and purposes.

Scientists and War

The relationship between scientific discovery, technological invention and war has varied, as even a quick-and-dirty history shows. In ancient times Archimedes refused to reveal the details of the engines of war he invented to defend his mother city, Syracuse, from the Romans. Earlier, the wheel was used on toys for hundreds of years before it was used on chariots. In China, gunpowder was for fireworks, not killing (Mumford, 1934, p. 84). War was only indirectly effected by discovery and innovation, except at those times when a new weapon or technology would revolutionize warfare completely. The Bronze Age became Iron as the stronger metal bested the earlier one. Chariots swept the ancient Middle East; Sun Tzu's subtle and codified insights into war allowed the State of Wu to dominate the Middle Kingdom; and English long bows cut down French knights in the Middle Ages. But after each innovation stability would quickly return. New methods often took hundreds of years before they were even applied to war.

In large part this is because war is such a strong cultural force. It often resisted the introduction of effective killing technologies for many years because they didn't fit the war discourse of the time. Even as recently as the seventeenth century in Japan, the warrior caste (samurai) succeeded in banning guns from 1637 to 1867 (Perrin, 1979).

Some have even argued that war and progress (technological and scientific) have been inexorably linked through the ages. There is much evidence to contradict this as a historical analysis, as John Nef shows in his book *War and Human Progress,* but the relationship of "progress" (techno-logical and scientific) and war today is almost symbiotic. Perhaps as many as half of all engineers and scientists in the United States work for the military and in accord with military priorities (Nef, 1963; D. Dickson, 1984). World-wide there are probably half a million scientists and engineers working on military problems (D. Dickson, 1984, p. 108).

The recent militarization of science is quite a change from the earlier relationship. Greek philosophy disdained war or anything practical. Roman engineering certainly focused on the needs of empire, but what we now call science (then still called philosophy) was hardly militarized. It was at the beginning of the modern era that this began to change significantly. The physics of cannonballs framed many of Galileo's questions, and the first major commercial product of optics was military telescopes.

Still, science was not focused on war even as war began to focus on science. Only 10 percent of the members of the early Royal Society did war work. The number is at least four times higher now (Ziman, 1976, p.11). And today's technoscience is much more powerful than the earlier alliance of engineering brotherhoods, mechanics, alchemists, natural philosophers, and mathematicians who founded modern science.

There have always been objections to linking what many described as a quest for knowledge with improving the killing of fellow humans. Although he served Cesare Borgia and later King Louis XII of France as a military engineer, Leonardo da Vinci explained that he wouldn't even describe his ideas for submarines in his secret notebooks because "of the evil nature of men, who would practice assassinations at the bottom of the seas by breaking the ships in their lowest parts and sinking them together with the crews who are in them" (quoted in Brodie, 1973, p. 239). Niccolò Tartaglia, the author of the first scientific study of gunnery, called simply enough *Nuova Scienzia* (1537), claimed that he was going to keep his work secret because "it was a thing blameworthy, shameful and barbarous, worthy of severe punishment before God and man, to wish to bring to perfection an art damageable to one's neighbor and destructive of the human race" (quoted in Brodie, 1973, p. 240). But later he changed his mind. He asserted that was because Italy was being menaced by Turkey, but why did he do the study in the first place?

John Napier, the Scottish inventor of logarithms in the early seventeenth century, feared a Catholic invasion of Britain, so he designed a tank and a gigantic cannon. However, he never revealed the details of his inventions because there were already too many "devices" for "the ruin and overthrow of man" (p. 243).

Even in 1945 such reservations were still heard, but to little effect.

Social Responsibility and Computer Scientists

Within weeks of the obliteration of Hiroshima, scientists from the Manhattan Project were organizing "to promote the attainment and use of scientific and technological advances in the best interest of humanity." They formed the Federation of Atomic Scientists (FAS) with other newly established scientists' groups and began publishing the *Bulletin of the Atomic Scientists* (Heims, 1980, p. 233). They also initiated the Pugwash Conferences, named after the town in Nova Scotia where they first met. In 1995 the Pugwash group was awarded the Nobel Peace Prize. But over the years scientists have been far from united.

The political divisions among physicists in the post-World War II period are well known: they ranged from Albert Einstein's pacifism to Edward Teller's militarism, and included the uproar over Oppenheimer's losing battle with McCarthyism. However, physics wasn't the only discipline where this drama took place. Computer science had its own martyr in Alan Turing, probably hounded to suicide by the British authorities because of his homosexuality.[1] And computer science has had its warlords as well, such as the analog expert Vannevar Bush, who directed the Office of Scientific Research and Development.

Bush is particularly interesting because he was directly and philosophically interested in the effect of science on war. He began his book *Modern Arms and Free Men*, an impassioned call for preparedness, with the statement that "Modern science has utterly changed the nature of war and is still changing it" (1949, pp. 2–3). Bush pointed out that it was Abraham Lincoln who first established official government science by setting up the Academy of Sciences during the Civil War. President Woodrow Wilson authorized the National Research Council for World War I. But in World War II with the National Defense Research Committee and the Office of Scientific Research and Development (employing over 30,000 people) a new level of science–war integration was achieved (p. 6).

This "new Era" as Bush calls it, actually started with World War I. He cites two factors: first, mass production and precision manufacture; second, the internal-combustion engine. "Between them," he says, "they made mechanized war possible, and the world will never be the same again." Submarines, airplanes, complex fire-control computing systems for gunnery on surface ships, and high-speed machine guns, all meant that war now required "a new kind of courage and endurance, a skill at operating machines under stress, and for the first time the factory behind the lines became a dominant element in the whole paraphernalia of war" (pp. 10–11). Two important concepts began here:

> First was the idea of reliable complexity in intricate devices, in masses of electrical and mechanical parts interconnected to function in a precisely predetermined manner, and dependable in spite of their intricacy because of the contribution of standardized mass-production methods. The second idea was that of relegating to a machine functions of computation and judgment formerly performed by men, because the machine could work more rapidly, more accurately, and more surely under stress. (p. 13)

Bush concludes, "When the First World War ended there were thus in existence nearly all the elements for scientific warfare." Also, in his view, "The long process of applying scientific results, all the way from the original academic theory or experiment to the finished device, had become ordered" (p. 16).

A key part of Bush's work, once scientists were integrated into policy making, was protecting young scientists from the actual fighting. In fact, he felt that "the problem of keeping young scientists in the laboratories was one of the toughest and most irritating problems we faced in the war" (p. 99). They needed to be protected because "armed prosperity" meant permanent mobilization. Scientists couldn't risk getting killed in battle because they were needed to constantly prepare for war:

> When war became total the armament race took on a new form. It is now a race not merely for the quick possession of a few battleships or other weapons that might decide an issue, but to attain immense strength of

every sort, such strength that the appalling costs of preparation can be paid without wrecking the system that produces them, such mounting strength that armed prosperity can proceed to more arms and more prosperity. (p. 120)

Bush became a central proponent of preparedness and of militarized science, although with many reservations. His was probably the mainstream response, but not all computer scientists agreed with him. Two of the most important "fathers" of computation, John von Neumann and Norbert Wiener, both of Hungarian–Jewish descent, took substantially different positions from Bush toward science and war, Wiener was almost a pacifist while von Neumann was particularly prowar. Comparing their postwar careers we can see how antiwar scientists have been marginalized politically even as their discoveries were being applied militarily whereas prowar scientists have been integrated into the highest levels of political power while claiming to be nothing more than objective experts without politics.[2]

When von Neumann was invited to join the FAS board he declined, claiming to have avoided "all participation in public activities, which are not of a purely technical nature." When invited again he wrote, "I do not want to appear in public in a not primarily technical context" (quoted in Heims, 1980, p. 235). This is the man who in arguing for the hydrogen bomb the year before had said, "I don't think any weapon can be too large" (p. 236). Even as early as 1946 he went out of his way to view the first postwar atomic tests, Operation Crossroad, which FAS objected strenuously to. He was hardly adverse to shaping military–political policy either: he met with Teller "immediately after they learned the news of the Soviet [atomic] bomb and discussed not whether but how to get political backing for an accelerated superbomb program" (pp. 240–247).

Norbert Wiener, the founder of the field of cybernetics, did substantial war work at MIT, but by the end of World War II he had decided that he would refrain from any further research with military applications. At the end of 1946 he said, "I do not expect to publish any future work of mine which may do damage in the hands of irresponsible militarists" (p. 208).

By 1950 von Neumann was advocating preventive war. "If you say why not bomb them tomorrow, I say why not today? If you say today at five o'clock, I say why not one o'clock?" (p. 247) In the early 1950s he chaired the committee that justified the development of intercontinental ballistic missiles (ICBMs), starting with the Atlas.

Steven Heims, whose fascinating double biography of Wiener and von Neumann is subtitled *From Mathematics to the Technologies of Life and Death*, concludes that this was a "deep failure":

On their own terms, the Atlas missile and its successors were greatly successful, like the hydrogen bomb before them. But when one recognizes, with Einstein, that the implication of the nuclear armaments race is that "in the end, there beckons more and more clearly general annihilation," this achievement on a deeper level represents a horrendous failure of Western civilization in its use of science and technology. (pp. 274–275)

Wiener's resistance to war science was as thoughtful as von Neumann's support of it was thoughtless. Wiener argued that science was a limited knowledge system that needed ethics and moral philosophy if it wasn't going to contribute to the destruction of humanity.

Yet, Heims reports:

In the years 1968–1972 I asked a considerable number of mathematicians and scientists about their opinions of Wiener's social concerns and his preoccupation with the uses of technology. The typical answer went something like this: "Wiener was a great mathematician, but he was also eccentric. When he began talking about society and the responsibility of scientists, a topic outside of his area of expertise, well I just couldn't take him seriously." (p. 343)

Von Neumann died young, of cancer, in great fear. A man who made death for so many possible became psychotic in the face of death. Wiener died old and happy. Perhaps their deaths don't reflect their lives, or maybe they do, but there certainly is evidence that their politics influenced their science. In his book Heims contrasts their science in crucial ways. According to Heims, von Neumann believed in mathematics totally; through the power of numbers and calculations he felt every important problem could be solved. Wiener argued that mathematics was limited, so he put forward ecological, interactive theories creating the discipline of cybernetics, and then spent his last years working on prosthetics and similar projects. The military has found many uses for both of their work. Of their politics and philosophies, von Neumann, who claimed to have none, became a powerful political figure while Wiener was marginalized. Von Neumann's politics of no politics was just right for postmodern war.

The reasons for this go right to the heart of science as we know it. Heims points to two "pillars" that "hold up the practice of science." The first pillar is "value neutrality" institutionalized, in his view, in the Royal Society. In return for royal protection and encouragement the society "outlawed the subjects of theology and politics from its meetings." The second pillar is the idea of the inevitable progress of science.

"In the seventeenth century," he goes on to point out:

the rhetoric of value neutrality tended to obscure the fact that the new science was the beginning of a radical subversion of the status quo through a scientific–technological–industrial revolution. In the twentieth century, however, the claim of value neutrality has . . . tended . . . to hide . . . [the

> fact] that scientific work has . . . strengthened the hand of the already dominant centers of economic or political power. (pp. 261–262)

Von Neumann agreed that science is neutral and that it is progress. These two claims are metarules, in that they pretend to be givens by the very definition of science most scientists ascribe to. Just as significantly, von Neumann was, according to Heims, "himself emotionally in tune with men of power." Hans Morgenthau, a friend of von Neumann, concluded in his own work on political systems that there is "in the great political masters a demonic and frantic striving for ever more power . . . which will be satisfied only when the last living man has been subjected to the master's will." He claimed this power lust was a form of love, and he saw this desire for power as intrinsic to humans and as the very basis for "power politics." (p. 323)

Some have argued that this "love" includes more base instincts among the individual scientists. As the repentant nuclear bomb designer Theodore Taylor admitted, "There is much to be learned by looking at the personal lives of the weaponers." He went on to elaborate on the intense personal pressures to develop new weapons: "There is the equivalent of an addiction . . . like drug abuse . . . power." In summarizing the relationship between science and the military he remarked, "We told the military what they wanted."[3] Many other observers have noticed this "hubris" factor in the technological push for new weapons (Broad, 1988; Vitale, 1985).

This dynamic has not changed. At ADS (Advanced Decision Systems), a military AI firm, a number of computer scientists admitted that they treated the generals and admirals they worked for like children. As Walter Bender, a scientist at MIT's Media Lab put it, "We cater to kindergarten children and admirals—people with very short attention spans" (Brand, 1987, p. 140). Other scientists, however, developed grave doubts. Susan Rosenbaum, for example, quit after ADS applied for a military contract she couldn't justify to herself:

> I had always rationalized it as, "Well, we only do defensive work; we're just defending the country." Then we started going after a military contract that was offensive. We didn't get it, ultimately, but just the decision to pursue that kind of work pushed me over the edge. I couldn't rationalize it anymore. (Faludi, 1987, p. 13)

Jeff Dan, another ex-ADS researcher, designed the software program PeaceNet that allows thousands of peace groups to communicate electronically around the world. He admits, "I'm kind of surprised more people haven't left ADS. Its depressing to think all those people are still there" (p. 13).

In the early 1980s the resistance within computer science to its own militarization led to the founding of Computer Professionals for Social Responsibility (CPSR), which has grown in the 1990s to dozens of chapters and thousands of members. CPSR led the critique of the SDI program and it has played a leading role in raising questions about military computing in general. CPSR has also been very concerned about the threat to privacy posed by the lack of constitutional protection for the information transmitted and stored via electronic media.

Most members of CPSR feel that science is a social process, albeit a very powerful one, and that scientists must take some responsibility for the social consequences of their work, just as Wiener and FAS argued. As scientists must take responsibility for what they do, they also must admit that they don't and can't know everything. The central argument against military computing, for example, is that computer science can never be perfected. It is ontologically and epistemologically limited and incomplete. It can never be as complex as nature, including human intelligence. This is certainly not a view shared by all scientists, but it is a guiding principle for some, such as Charles Schwartz, a physics professor at the University of California at Berkeley, who refuses to teach graduate students physics anymore (Schwartz, 1988). Or Prof. Joseph Weizenbaum of MIT, who has called on computer scientists not to do vision research since it is framed by military priorities. There are also less established scientists, such as a recent Ph.D. in AI from the University of California who refuses to publish any of her research because the military could use it. In 1995, the student Pugwash movement, an offshoot of the scientist's group founded after World War II, started circulating a pledge among science students to forgo war work.

On the other hand, mainstream science, with its emphasis on experimentation and materialism, is still generally oblivious to its role in postmodern war. As powerful as establishment science is, it is important to note that alternative visions of science continue to flourish in many surprising places, sustained by the obvious contradictions of postmodern science.

These contradictions have had an equal impact in a very surprising place—the military. Just as many scientists are trying to redefine what a scientist is, so are many military people redefining their own role.

Soldiers and Generals for the End of War

> Abolish war in all its forms!
> —*Position statement of Veterans for Peace*

Since the first days of modern war military men and others have been arguing that new weapons would make war impossible. The poet John Donne was one of many who claimed that artillery made war too horrible to fight (Brodie and Brodie, 1973, p. 70). Benjamin Franklin, on seeing the first balloon flight, predicted that balloons might make war obsolete (Sherry, 1987, p. 92). John Hay, Abraham Lincoln's personal secretary and later Theodore Roosevelt's secretary of state, assumed it to be "the plain lesson of history that the periods of peace have been longer protracted as the cost of destructiveness of war have increased" (Sherry, 1987, p. 92). Victor Hugo had a similar prediction regarding flying machines. He thought they would make armies "vanish, and with them the whole business of war, exploitation and subjugation" (Clarke, 1966, p. 3). Jack London said that "the marvelous and awful machinery of warfare . . . today defeats its own end. Made preeminently to kill, its chief effect is to make killing quite the unusual thing" (Sherry, 1987, p. 92). This has been quite the popular theory among inventors: Robert Fulton, Alfred Nobel, and Thomas A. Edison are among those who assumed their new technologies would make war unfightable (C. H. Gray, 1994). Henry Adams and H. G. Wells predicted either human suicide or the end of war. As R. Buckminster Fuller put it, "Either war is obsolete, or we are" (Sherry, 1987, p. 92). But why hasn't everyone accepted that war is obsolete?

It is true, and important, that war is no longer universally considered necessary, inevitable, or even good, as John Mueller notes:

> Many of the most fervent war supporters seemed beyond logical or practical appeal because they were so intensely romantic about their subject. Others were attracted to war because they believed it to be beneficial and progressive, and many, including some who loathed war, considered it to be natural and inevitable. Most of these views, particularly the romantic ones, were encouraged by the widespread assumption that war in the developed world would be short and cheap. . . . None of these lines of thinking has serious advocates today, particularly as far as they pertain to international war in the developed world. (1989, p. 39)

Mueller is too optimistic. War obviously has its supporters to this day, but now they are on the defensive. While there have always been those who argued war was insane, lately the number has been growing. During World War I this became a mass movement among soldiers. The British antiwar soldier poets and the Russian, French, and German mutinies were all reflections of the terrible lesson of trench warfare: war is now inhuman.

This idea became broadly held in the United States during the Vietnam War. Widespread draft resistance, desertion, and mutinies in the field, at sea, and even in Guam and California shook the U.S. military. Only American English has a special word, "fragging" (from fragmentation grenades), for soldiers' killing of their own officers. Once back in "the World" (the United States) vets formed a number of antiwar groups, most notably Vietnam Veterans Against the War (VVAW). Today such activ-

ism continues within the mainstream veterans organizations and in special veterans groups, such as the Veteran Action Teams that have gone
to Central America and the Veterans' Peace Convoys that managed to
take humanitarian aid to Nicaragua despite U.S. government obstruc-
tions. In the early 1990s, veterans from the United States and Canada
formed a group to abolish war by the year 2000. There are also several
important peace-oriented think tanks founded by retired military men.
The most important is the Center for Defense Information, whose leaders
have included Adm. Gene La Roque, Rear Adm. Eugene Carroll, Jr.,
Capt. James Bush, Col. James Donovan, and Maj. Gen. Jack Kidd. There
is also the Institute for Space and Security Studies, founded by Lt. Col.
Robert Bowman, who once was in charge of military space research for
the Air Force. Even a dozen NATO and Warsaw Treaty Organization
generals met and drafted statements saying that soldiers now must prevent
war and that the first principle of military virtue is "to serve peace." They
claim high technology has made war impossible (Generals [Brig. M.
Harbottle et al.], 1984, p. 23): "In a nuclear age, the soldier's commitment
can only be to the prevention of war. The first principle of military virtues
must be therefore: 'A soldier has to serve peace' " (p. 23).

Many other military leaders have reached the same conclusion, including Dwight D. Eisenhower, Adm. Lord Louis Mountbattan, and Gen. Omar
Bradley. Gen. Douglas MacArthur, of all people, said on September 2, 1945,
after accepting the Japanese surrender that

> Military alliances, balances of power, leagues of nations, all in turn failed,
> leaving the only path to be by way of the crucible of war. The utter
> destructiveness of war now blocks out this alternative. We have had our
> last chance. If we will not devise some greater and more equitable solution,
> Armageddon will be at our door. (quoted in Meistrich, 1991, p. 102)

Of course, a few years later he was demanding the right to invade China.
This follows a familiar pattern. Commanders at war seldom see war as
impossible. Adm. Hyman Rickover, the "father" of the U.S. nuclear Navy,
served until he was over 80 years old. Soon after retirement he denounced
his own career. As with the vast majority of the military repentants, he waited
until his career was finished to renounce war and preparation for war.
John F. Kennedy warned that "Mankind must put an end to war or war will
put an end to mankind," a year later the world was brought to the brink during
the Cuban missile crisis (p. 102).

Dissent is, of course, only a part of postmodern military culture. The
dominant discourse is still controlled by believers in war. They will find wars
or make them. They will also rearticulate the military's role so as to recoup
the fear of total war into more military power. The slogan of the Strategic

Air Command (Minuteman and MX missiles; B-52 and B-1 bombers) is "Peace is our Profession." The justification for Star Wars is framed in the same way, as were the invasion of Grenada, the kidnapping of Noriega (in the search for drug "peace"), and the war for Kuwait.

The recent upsurge in UN and NATO peace missions discussed in Chapter 1 represents something new, both qualitatively and quantitatively. Whether or not it means the military can convert to peacekeeping we don't know yet, but the growing disatisfaction with war has clearly spread to the warriors themselves, and that is reason for hope.

But even as military culture struggles with whether or not war will continue, others from outside have sought to redefine what being a warrior means as well.

New Types of Warriors

This very refusal of war might be the next step in the evolution of warriors. It is not a new idea. Through the ages soldiers have become nonviolent, and nonviolent activists such as Mohandas K. Gandhi have called for pacifist warriors. This can lead to bitter irony.

Brian Wilson, who served as an Air Force intelligence officer in Vietnam, later lost his legs trying to block a weapons shipment from the Concord Naval Air Station in California to Central America. Being a peace activist was much more dangerous to Wilson than serving as a soldier.

This is a paradox. Many peace activists around the world, even in Western countries, have had friends murdered by the state; they have seen people killed; they have been chased by cars, trucks, police, soldiers; and they have been captured, beaten, and locked up. These nonviolent activists have seen more violence than most military people. Their emotional experience can be quite similar to men and women in combat. Only their refusal to kill makes them different from warriors in the commonly accepted sense.

So, unsurprisingly, there are also strong currents in the peace movement proper that are trying to remake the warrior metaphor and claim it for themselves. Sometimes they look back to premodern versions of war. Native Americans and their supporters have often referred to themselves as rainbow warriors, a title also taken up by the Greenpeace eco-activists. Within grass-roots antiwar and peace groups there have been proposals for activists to make a commitment of several years to a peace army (from the First Strike Project)[4] or to work for peace (Beyond War). Established pacifists have an international peace brigade program that inserts peace "soldiers" into war zones as varied as Guatemala, Kuwait, and Sri Lanka, as the religious Witness for Peace and the Veterans Action Teams are doing in Latin America.

While most peace people, as they often call themselves, still manifest an open aversion to militarism and military metaphors, the influence of the warrior image is quite powerful, being used by draft resisters, veterans, feminists, Christians, and non-Christians. Even some neopagan peace activists are putting forward "The Path of the Pagan Warrior as Revolutionary Activism Today . . . [in] . . . Defense of Mother Earth & the Web of Life."

This highlights an intriguing aspect of this development: these peaceful warriors are quite often women. Of course, in the official U.S. military many of the soldiers are also women, although strict (and ineffective) divisions are made between warrior specialties and technical ones, and women are underrepresented in the higher ranks. In the peace movement it is almost the reverse. A majority of activists are women, as is an even larger proportion of the leadership.[5] In the mid-1990s women took the lead in waging peace is Yugoslavia, Angola, Russia, Nicaragua, Ireland, the Philippines. It is a social phenomena even Pope John Paul II has noticed (Lederer, 1995).

So the very definitions of science and war, scientists and warriors, are being contested. If postmodernism means anything in relation to war and science it is the inevitability of change. Modern war was certainly never stable as an institution, but before it became suicidal it at least made limited sense. Postmodern war is much less sensible.

War has become so dangerous it threatens everything. Perhaps it can be ended or changed enough so we can survive. What it means to the future/present is in part already determined, reified in institutions, fixed in artifacts, and inscribed on our bodies. But there is that other part that has yet to be formed. Empires fall. Icons are shattered and reformed. Definitions change. New metaphors are forged. French philosopher Michel Foucault pointed out how important this can be:

> The successes of history belong to those who are capable of seizing these rules, to replace those who had used them, to disguise themselves so as to pervert them, invert their meanings, and redirect them against those who had initially imposed them; controlling this complex mechanism, they will make it function so as to overcome the rulers through their own rules. (1977, p. 86)

Foucault claims we can always have a say, even if it is disguised, perverted, or inverted at times. And to such deceptions go "the successes of history." True as this is, some things are better not said, or done, at all. The ends are not separate from the means. Which must serve as a warning to anyone who wants to co-opt one of the central tropes of the informatics of domination, the warrior, or the idea of pure information. Still, there do seem to be real possibilities in rejecting the role of victim (of science, of the warrior ethos) to embrace instead whatever moral possibilities there are to be active shapers of these cultural constructions. Perhaps deconstructing the warrior ethos and inverting it can be complementary. Maybe we don't have to do away totally with warriors (and how can we resist being cyborgs?), but cyborg-soldiers, autonomous weapons, war managers, and technobureaucrats seem much less inevitable.

These are times of great change, so there are possibilities. They are just not simple.

Chapter Thirteen

War and Peace 2000

Between Horrific Apocalypse and Utopian Dream

Postmodern war seems to be coming to an end yet to be determined. Currently, war's prospects oscillate between two very different poles: some form of horrific apocalypse or the even less likely possibility of general disarmament and peace.[1] A dynamic fluctuation between them might mean a limited nuclear war or biological attack, with deaths in the millions, that would precipitate a general reevaluation of political realities. Without that "shock treatment" scenario we can expect a low level of continual horrors leading (at best) gradually to the kind of fundamental changes necessary for humanity to survive. It is almost a choice between a postwar world and a postcivilization one. However the conversation, the discourse of postmodern war goes, we will all have some say.

Postmodern war is a discourse for two reasons. *First, its unity is rhetorical.* It is by virtue of the metaphors and symbols that structure it, not by any direct continuity of weapons, tactics, or strategy between its various manifestations. A central metarule is "change." Once postmodern war ceases its incessant mutation it will no longer be postmodern war, and it may not be war at all. *Second, it can be described as a culture.* Ideas, material artifacts (especially weapons), people, and institutions are all jumbled together. The working out of what actually happens to this jumble very much resembles a conversation, all the more so since much, but certainly not all, of what happens is done with words, either written or spoken, and since the main actors (though not the only ones) and most of the judges (especially of definitions) are humans.

Looking at war as a discourse helps explain some of postmodern war's novel features. The most important of these is the centrality of information to power and rhetoric to knowledge. A discourse analysis also points to why shifting rules about bodies–machines and about gender are so crucial to understanding postmodern war. Finally, the discourse metaphor helps reveal some things about the possible future of war that are not readily apparent, such as illusions of control that might lead to apocalypse, or the restructuring of military missions that validate peacemaking.

Information, Rhetoric, Power, Knowledge

In an analysis of power Michel Foucault quotes J. M. Servan, who in 1767 wrote a "Discourse on the Administration of Criminal Justice":

> A stupid despot may constrain his slaves with iron chains; but a true politician binds them even more strongly by the chain of their own ideas; it is at the stable point of reason that he secures the end of the chain; this link is all the stronger in that we do not know of what it is made and we believe it to be our own work; despair and time eat away the bonds of iron and steel, but they are powerless against the habitual union of ideas, they can only tighten it still more; and on the soft fibers of the brain is founded the unshakable base of the soundest of Empires. (quoted in Foucault, 1977, pp. 102–103)

Servan was arguing for the forging of a link between the ideas of crime and punishment; much as links have been forged in the discourse of the U.S. military between high technology and victory and between scientific progress and military superiority. Foucault saw this form of "ideological power" as being superseded, or at least supplemented, in the last few hundred years by a new "political anatomy," which focused on the body as the ultimate object to be controlled, and made up of an economy of "punishments" and rewards. Cutting off the hands of thieves or torturing criminals and dissidents or mutilating their corpses gave way in much of the world to scientific prisons and criminal psychology. Today, it is the guardian of democracy, the United States of America, that has the highest percentage of its own citizens incarcerated.

In a similar way the discourse of the postmodern military is based not just on chained logics of "information = victory" and "more bombs = more safety" and the traditional "enemy = evil" and "us = good." There is also an array of particular subthemes (cyborgism, epistemophilia, technism), each one a discourse in its own right, that elaborate on these links and form a working economy of inducements and punishments that insist things must be the way they are.

This economy is not just the main discourse logic and the explicit and implicit arguments outside the central themes, such as the claims that the Strategic Computing Program (SCP) may not be good military policy but is great economic policy (for competing with the Japanese) or great scientific policy (for the love of knowing). There are reasons and connections that work on deeper levels. Beneath the rule that says "high technology equals military power" are the meanings of "high technology" and "military power" them-selves—meanings in the sense of the metaphors and other tropes that form the dictionary and speaking meanings, and meanings in the sense of the emotions triggered by the terms themselves and their associated concepts.

In the discourse system, the connections between subjects, institutions, artifacts, and actual discourses are usually unexamined metarules. One of the most important of these is metaphor. It is clear now that metaphors (and all powerful tropes) are often metarules. George Lakoff and Mark Johnson point out in their book *Metaphors We Live By* that the "essence of metaphor is understanding and experiencing one kind of thing in terms of another." They go on and argue that if one defines metaphor broadly, then "human thought processes are largely metaphorical" (1980, pp. 5–6). Therefore, the metaphors we use to communicate our thoughts often play a major role in structuring them as well. That is one important reason to pay close attention to the dominant metaphors of postmodern war. Metaphors also delimit in many ways what we can think and feel. This is just as important a reason to pay attention to them.

Looking at the genealogy of postmodern war, it seems that changing the metaphors is one way of changing the metarules. The central metarule of war is its currency, human bodies. As they change in war, so changes war.

Bodies, Machines, Cyborgs, Genders

> The man–machine system is more perfect than man or machine alone.
> —*Druzhinin and Kontorov (1975, p. 17)*

It is actual bodies that are shaped by war. Writing about the Freikorps soldiers, the right-wing World War I German veterans who crushed the left in the early 1920s and fostered Nazism at the end of that decade, Klaus Theweleit remarks:

> The relationship of human bodies to the larger world of objective reality grows out of one's relationship to one's own body and to other human bodies. The relationship of the larger world in turn determines the way in which these bodies speak of themselves, of objects, and of relationships to objects. (1987, Vol. 1, p. 24)

Theweleit shows how the war cyborg became a dominant type in militaristic discourse after World War I. The rejection of the body that he notes in detail in the lives and writing of German protofascists is not a necessary part of cyborgization, but it is certainly a real temptation for many warriors. Even without the total rejection of the human, the line between the human and the machine, body and object, has never been so vague as it is in contemporary man–machine weapon systems.

Outlandish as most of these military projects are, and strange as the concerns of philosophers might seem with our talk of the end of war and cyborg identities, it is important to remember that all this talk is based on a very mundane reality (when seen up close) of tax dollars, tens of millions of soldiers, millions of war machines, countless computer programs, and hundreds of thousands of working scientists.

It does not seem accidental that as the line between human and machine becomes vague there is growing confusion on the differences between male and female as well, especially in their relationship to war. Women have always been crucial to war, but it has seldom been admitted, by militarists or pacifists. Cynthia Enloe is a notable exception:

> In each country military strategists *need* women. They need women who will act and think as patriarchy expects women to act and think. And they need women whose use can be disguised, so that the military can remain the quintessentially "masculine" institution, the bastion of "manliness." (1983, p. 220; emphasis in original)

But women's role is hidden no longer; now it is highlighted. Why should this be? A crucial factor in understanding the rules and metarules of any system of power/knowledge is to know what other discourse systems it is entangled with. In the case of postmodern war, already the intersection of science and war, there have been many signs that gender is also key. Science and war are clearly gendered as masculine domains. Within both, many subjugated knowledges take on alternative gender markings—either explicitly, as in the case of feminist critiques of science and war, or implicitly, as when emotion is counterposed to rationality. Catherine Lutz describes in her work how permanent war, based in part on "the massive masculinity of the military," produces the contemporary psychological ethic (1995).

So it isn't surprising that computer weapon systems are clearly gendered as well:

> They are "masculine," in the full ideological sense of that word which includes, integrally, soldiering, and violence. There is nothing far-fetched in the suggestion that much AI research reflects a social relationship: "intelligent" behavior means the instrumental power Western "man" has developed to an unprecedented extent under capitalism and which he has always wielded over woman. (Edwards, 1986b, p. 40)

But the dynamic is actually a little more complex. Yes, the powerful machines are gendered male in most respects, but now the soldiers are less clearly defined. AI does seem to represent the social relations of postindustrial societies—as the machines are construed as masculine, the humans become feminized or even neutered, especially the lower ranked worker-soldiers. In the context of postmodern war, this allows for the integration of woman soldiers into the military, which is crucial for practical reasons (their skills are needed) and social reasons (it is their right). They take on some traits gendered masculine; at the same time their fellow male soldiers seem "feminine," subordinated to officers and to powerful killing machines in weapon systems. While such coding of lower ranks as feminine has always existed to some extent, the traditional masculine identity of soldiers as killers used to dominate. Now the gender, as well as the sex, of soldiers is much more ambiguous.

It is everyone's work, by and large, to make these killing machines. The vast high-tech apparatus of postmodern war is, in terms of size, potency, and projective power, the greatest work of humans ever. But it is a project that is managed, overwhelmingly, by men, and many of them know this and are proud of it. It is not an act of creation that is well regarded under many other discourse systems; but in most ways all other discourse systems are subordinate to those that frame and maintain postmodern war. This is because the system of systems, the metadiscourse that makes up human culture, is held hostage by war in a fundamentally new way. Now war is consuming the rest of culture through its ubiquitous discourse, destroying it as truly as any directly technoscientific wargasm might. Why is this?

The Causes of Postmodern War

The two most important causes of war that are usually ignored in war's own discourse seem to be (1) the character of technoscience and (2) the importance of emotions in how humans relate to technoscience and war. Greed (economics), ambition (human nature), and stupidity certainly play an important role in causing wars, but they don't explain why war should continue when it threatens the human future.

Technoscience calls war into question (war will destroy the world) and simultaneously provides the rationales for continuing it (war can now be managed; war can be fought between bloodless machines). High-tech weapons, especially nuclear weapons, have also shifted the discourse of war in some crucial ways—freezing out world war and total industrial war, except in the budgets and imaginations of the armies of the world, and fostering dreams of painless "surgical" LICs.

Postmodern war exists because it has managed to deploy a specific and limited definition of rationality and science as an institution to replace valor. The assumed rules of this system are arguments of cost–benefit analysis, taken from business management, from science and technology, and various other types of sharply defined and constrained rationality. The values are those of lethality (killing humans and destroying machines) and low casualties for "our side." The rhetoric pits the human enemies' (the other) machines against "our" good technoscience. Patriotic rhetoric has become the rhetoric of technological power. Consider the spectacle of jets at air shows in tight formation, of grand shuttle launches and landings, of proud parades and public rituals welcoming the new weapons systems into the military fold. The celebrations become even more intense when the weapons are actually used, witness the technoeuphoria of the 1991–1992 Gulf War.

There is also a pronounced bureaucratic structure to technoscience's character that is especially visible in postmodern war. Implicitly military technoscience, like all technoscience, is in many ways a bureaucracy (and a network of bureaucracies), with interests in perpetuating itself in the realm of budgets, projects, and so on. It is made up of the bureaucracies of the U.S. Congress, the White House, the computing and other companies, the research facilities, the military services and commands such as ARPA, and the standing forces themselves. Bureaucracies always hide behind their explicit rules and run on their implicit standards toward their unstated, obvious, goals of status, advancement, perpetuation, and power. These goals are as emotional as they are practical. This book has shown that conscious desires are augmented by unconscious impulses to cope with the extraordinary anxieties of postmodern war, to transcend human limitations such as corporeality and mortality, and to continue the satisfying masculine rituals of war.

In a very real sense technoscience is being developed *unconsciously*. Consciously each project is just another weapon system, another source of contracts, grants, and eventually military power; unconsciously such projects offer life (battle) without death, erotic machines in place of bodies, and war eternal. For humans to think of controlling postmodern war we must understand the various forces that drive it.

Geoffrey Blainey shows in *The Causes of War* (1973) that a key element in the vast majority of wars is the serious misperception by both sides that they could probably win what they needed from the war. Perception is largely psychological of course, despite its physiological basis. The emotional attraction technophilia gives high-tech weapons also means that those who feel it overestimate the power of these weapons. Belief in these weapons leads to the kinds of misperceptions that have caused most wars. Belief in high-tech weapons leads to more wars.

Many of the researchers discussed in this book point out that the relationship between humans and technoscience is an emotional one. Zoë Sofia calls it epistemophilia (1983); Robert Jay Lifton (1987) labels it technism and nuclearism; Langdon Winner (1986) has his technological somnambulism, and Hans Morgenthau (1962) speaks openly of power lust. In different ways they all argue that technoscience has been politicized, even eroticized. What is crucial isn't this political and erotic content, however, although technology has never been so intensely eroticized and politicized as it is in postmodern war. What is important is that the potent emotional seduction of postmodern war is being disguised as rationality, especially to its most fervent admirers.

They don't just like technoscience and appreciate its usefulness, they *love* it. It is connected to their sense of power intimately. Consciously they expect, with the assistance of science and technology, to move mountains, speed to distant friends, watch pictures from the moon, the Amazon, and London, see bodies play and fight in slow motion, live longer, and smite their attackers. Unconsciously it goes even further. They dream they can create life and live forever. They fantasize that they can totally dominate the other—be it a human or the earth itself.

The fear that such power evokes can be blunted by denial or through doubling, the creation of emotionally immune "other" selves that are professional or cynical or ignorant. Or it can be dealt with, as Sofia, Carol Cohn, and Lifton seem to agree it often is, by turning it into a love for what is feared. The people who plan, build, and use these war systems are exactly the people who, day after day, think about and plan war. They live in a world of war where war becomes the best solution for every problem, including the problem of war itself. This isn't obvious to outsiders because they aren't considered "experts" on war. But it is the experts who have been seduced. It is the experts who have lost their judgment because of the very intimacy with war that their expertise is based on. The implication is that this cycle must be broken, internally if possible, externally if necessary, if war is to be understood instead of worshipped.

Whatever other factors are involved, it seems significant that so much emotion can be easily uncovered within this power–knowledge system that lays such strong claims to rationality, transparent rhetoric, and value-free logic. What is portrayed as a disengagement of the emotional from the rational actually seems more of a disengagement from embodiment. Thus cyborgization is ignored, because to notice it would be to notice the human body and how it is being changed. Meanwhile, the emotional is not actually disengaged so much as it is displaced in a number of different ways. It is turned back, in the form of erotic attachment and gratitude, onto the technologies and weapons that generate fear in the first place. As fear and loathing it is placed onto the *others* (the evil communists or Hitler-like enemy leaders or drug lords, on the logical level, and women and nature, on a more latent level). It is even displaced onto these high-tech war machines, granting them rhetorical embodiment and other human qualities even as the humans are portrayed as emotionless and are thus reduced to "thinking heads."

Certainly in this discourse, on a rhetorical level, there is the displacement of responsibility, intentionality, even embodiedness. These qualities are transferred through metaphor and other figures of speech and image from the human to the machines. Then these machines become the subject of the text in the same way weapons are the subject of strategic discourse. Bringing the machines into existence is the whole reason for these texts and projects, as strategic discourse is built around what weapons are needed for security and the way these weapons might be used. Nuclear weapons, for example, are a necessity of strategic discourse specifically, as computer weapons and applications generally are seen as an undeniable necessity for the whole Western style of postmodern war.

This system seems to feed off the emotions it represses. Rational war becomes the controlled hysteria of nuclearism, and then it threatens to become a technological Thanatos, the death instinct Sigmund Freud brooded over so much at the end of his days. When the roots of war's ritual attraction are considered, the connection of destruction to creation in postmodern war makes at least psychopoetic and mythological sense. What Susan Mansfield said about rituals certainly applies to war: "Two themes are universally present in the ritual: death and fertility. Death dominates the initial and central steps; fertility is a subtheme that emerges to dominate the conclusion" (Mansfield, 1982, p. 28).

Death and fertility: postmodern war and its proliferation of military technology; death beyond imagining and technoscience procreation, man the father and mother both. Subconsciously the vast destruction of war implies an equally vast fertility. Unfortunately, postmodern war only breeds new weapons and new wars. As Zoë Sofia shows in her work (1983, 1984), it is an emotional economy that is quite patriarchal: the subject is always masculine, even when it is a machine; as the object is always feminine, even if it is a 19-year-old soldier-boy. It is mediated, in popular culture, through aliens, women combat leaders, cyborgs, and intelligent computers. In war it is reflected in the shifting status of humans and machines and the intense repression of intense emotions. It is described by unbelievers, politically and rhetorically, as ritualized ideology, the ideology of preparedness, digital realism, magic bulletism, technism, nuclearism, epistemophilia, technophilia, short-term and vulgar rationality, mechanized thinking, and technological fanaticism. For believers it is only natural.

Considering these powerful underlying emotional dynamics it is no surprise that, to a large extent, the war managers only manage the system physically and politically; psychologically it manages them. It generates emotional needs and meets them. It offers power over death through megadeath and artificial life. It offers the illusion of control that will only dissolve for some under the hard rain of real war. It is a complete emotional universe, as long as there is war and rumors of war and therefore preparing for war. This is for the managers. For the soldiers, and for the civilians who are now, all of us, the prime targets of nuclear weapons, smart bombs, terrorists, or drug raids, it is a different matter. Most humans don't have an economic or a psychological investment in war.

In concluding his article on the AirLand Battle manager, Gary Chapman, who served as a sergeant with the Green Berets in Vietnam, comments:

> We might like to be able to "manage" the living hell that modern battle is likely to produce, but the chances of achieving that seem remote. We have decided to build military systems that outstrip our ability to understand war, which has always been the most chaotic and least understood human activity. Automating our ignorance of how to cope with war will produce only more disaster. (1987a, p. 7)

Where we don't understand but pretend we do, where we refuse to acknowledge that we do indeed feel, there and then emotions take command. To challenge that command we must take political action, for in the realm of political psychology there is no effective therapy, only policy. Setting that policy is not a psychological act but a political one. Postmodern war is first of all, and last of all, a political issue.

Changing the Future of Postmodern War

Never before have so many people assumed that war as an institution has outlived its usefulness. John Keegan, at the end of his brilliant evocation of soldiers' actual experience of the battles of Waterloo and the Somme, *The Face of Battle*, notes, "the usefulness of future battle is widely doubted." He goes on say, "It remains for armies to admit that the battles of the future will be fought in never-never land." Even though soldiers continue to "show each other the iron face of war . . . the suspicion grows that battle has already abolished itself" (1976, p. 343). In *A History of Warfare* he is even stronger, declaring that war cannot continue (1993, p. 391).

Technology has rendered war too dangerous, leading to changes in the discourse of war:

> Where war was once accepted as inevitably a part of the human condition, regrettable in its tragic details but offering valued compensations in opportunities for valor and for human greatness—or, more recently, in opportunities for the ascendancy of superior peoples—the modern attitude has moved towards rejection of the concept of war as a means of resolving international or other disputes. (Brodie, 1973, p. 274)

Brodie goes on to add that it is "especially striking" that the only acceptable argument for war now seems to be self-defense, "expanded by the superpowers to include defense of client states" and only in a few cases the "correction . . . of blatant injustice." This may seem a cosmetic change, for in "self-defense" and "justice" there is room to justify almost every war. But in terms of war as a discourse it is an important distinction.

War discourse now is dependent on the desirability of peace to justify war. It is an important contradiction. The claim that war is natural was perhaps the central assumption of almost all war discourse up until the twentieth century. Most "just war" theories assumed that, as some wars were just, they were therefore natural. Then war began to assume unnatural proportions, and for many people pacifism began to seem natural. Still, traditional war discourse, tied in closely with ideas of masculinity and self-worth, not to mention the physical consequences of losing as opposed to winning wars, is still being mobilized. Now the older images are being deployed along with the "peace through war" trope that George Orwell found so ironic ("war is peace"), to make sure war isn't abolished or that it doesn't wither away. And just as importantly, postmodern war is now framed through computer metaphors and weapons as a manageable contest of intelligent machines in cyberspace, making it a much less horrible prospect. So, in spite of the invention and the use of the most incredible inhuman weapons the danger and reality of horrible wars remains at the same extraordinary level that it has since 1939.

This is one of the central lessons of this story—*war discourse is tenacious*. It is deeply embedded in human culture in many different ways. Remember, war is older than capitalism, older than socialism. It is older than civilization. War has a life of its own. In Barbara Ehrenreich's maxim, one of the main reasons we still have war is "War for war's sake" (1991, p. 23). Discourses, like bureaucracies, are self-perpetuating. This said it is important to reiterate that war is not the same as human culture. There has always been antiwar sentiment as well. Most people who have ever lived never fought in wars of any kind. And, as this book shows, war has only survived by reconfiguring and reinventing itself. War may be an old and powerful discourse, but it is a limited discourse all the same. It can be eradicated from human culture, but only if it is challenged as war per se, not just in its current postmodern incarnation. War's centrality to masculinity, its claims to spiritual power, and its defining role in politics have to be confronted specifically or the discussion will just shift from its obviously insane aspects. Some say war can never be abolished, but the same was said about slavery. It to was a very old discourse, and yet now it is almost totally discredited. Abolition of war is possible. Lately, it has even seemed very possible in the abstract but not in practice.

Unfortunately, the collapse of the Warsaw Pact and the Soviet Union does not mean peace is about to break out. On the contrary, the United States sees Russian decline as a green light for more military adventures or for even official declarations of *pax Americana*. In February 1992 it was revealed in leaked Pentagon documents (Tyler, 1992) that the DoD was planning for a strategy aimed at keeping any other power from even playing a regional role in the world. Europe and Japan were mentioned as particularly dangerous potential rivals. Possible midintensity wars were foreseen in Cuba, Iraq, Korea, the Indian subcontinent, China, and the old Soviet Union, echoing the military's shift to mid- and low-intensity conflict that dates from at least 1988 (Klare, 1990). The one-superpower world justifies a $6 trillion spending policy for the U.S. military that would leave it with 1.6 million soldiers and unprecedented world military dominance.

Meanwhile, the old Soviet Union has already become a site for war itself, and in general the other schism's of the world, especially between poor and rich, remain as deep as ever. With the breakdown of the bipolar world order, the danger of regional conflicts and "small" nuclear wars increases (Ramezoni, 1989). The Soviet–U.S. Cold War has recently been called a "Cruel Peace" (Inglis, 1992). The contemporary "peace" of *pax Americana* is certainly no kinder. In the next few decades U.S. military interventions will no doubt continue, probably until some defeat convinces the government that war is still dangerous and unpredictable. Despite the *pax Americana* the chances of a horrible (high-tech, high-casualty) regional war have also increased in the new multipolar world, especially in the Middle East.

On the level of LIC, the replacement of men by machines with the hope that casualties will be kept down makes wars like Vietnam, Afghanistan, and those in Central America much more politically possible for the West. Illusions about managing war and the precise application of force in the form of jet bombers with smart bombs or the battleship USS *New Jersey* have also contributed to a number of bloody low-intensity adventures by the United States in Lebanon, Libya, Grenada, Panama, Somalia, Haiti, and Bosnia–Herzegovina. Nothing has happened to dispel these illusions. Individual Western leaders will now be more tempted than ever to displace bad economic news or dropping popularity polls with raids on demonized nations, especially during election years.

When both sides have high-tech weapons they believe in, which could well be often since they are sold around the world, midintensity conflicts can easily result. Without the superpower rivalry more regional wars will break out between traditional enemies, and there will continue to be civil wars as well. Some of these conflicts could become small nuclear wars, if there is such a thing. Perhaps it makes sense to call them limited nuclear wars, since the number of weapons used would be limited and would probably stop below the threshold of global annihilation, although they may cause apocalyptic changes to the world's climate. The continued proliferation of nuclear weapons, as well as the spread of computer command and control, means that the chance that total war could start accidentally or escalate from a limited conflict remains as high as ever if it isn't getting worse.

Three Russian analysts, at a Sipri conference in Sweden on weapons and AI in 1986, predicted that the growing importance of computers in war meant that it was very possible to start a "new round of the arms race based in large part on the military application of new information technology" (Kochetkov et al., 1987). This is exactly what many government leaders within the United States have called for, although with a focus on LIC under its latest label, cyberwar.

Another important way the high-tech weaponry can negate nuclear disarmament is also discussed in the Russians' paper. New information technology, with the use of rapidly improving conventional explosives, has the potential of greatly increasing the lethality of the battlefield: "Thus a situation is created in which increasing the accuracy of conventional weapons becomes equivalent to increasing their explosive power up to the level of nuclear weapons" (p. 160). Hence, the United States can plan to win midintensity conflicts by deploying conventional weapons with the power of small nuclear ones, unless their opponent has similar weapons or biochemical weapons with equal effect.

As if this were not bizarre enough, Marvin Minsky has pointed to another problem military computing might lead to—insane fighting machines:

> The first AI system of large capability will have many layers of poorly understood control structure and obscurely encoded "knowledge." There are serious problems about such a machine's goal structure: If it cannot edit its high level intentions, it may not be smart enough to be useful, but if it can, how can the designers anticipate the machine it evolves into? In a word, I would expect the first self-improving AI machines to become "psychotic" in many ways, and it may take many generations . . . to "stabilize" them. (quoted in Forester, 1985, p. 562)

So it is important not just to reduce the level of nuclear weapons but also to control the military use of high technology, in general, and information technology, in particular, or nuclear disarmament will be meaningless.

One of the other main themes to emerge in postmodern war is that technologies have unbalanced the distinction in war between the rational and the emotional. While this is in many ways a false dichotomy (and not one of my creation), it is a central image of war discourse and one that has restrained war somewhat. As long as it was accepted that war was in many ways irrational it could never be assumed that it was predictable and controllable. Pretending that emotion has been eliminated from war is a necessary precondition for the illusion that war is nothing more than a particularly severe form of implementing a rational policy. So how can the discourse be changed?

Deconstructing War

While war discourse is certainly not totally under our control, we humans can intervene in the system physically or verbally in ways that help restructure that very discourse. For example, war is not peace, it is not machines that die in war but people, and the real motivations for war are never the official ones. Just explaining these elements of postmodern war deconstructs much of its justification, if the explanations are believed and taken to heart. Such an intervention can be thought of as changing the rules of the conversation or, better yet, the metarules of war's discourse. This book has pointed to four basic ways these rules about the rules might be reconfigured:

1. Actual physical constructions such as machine guns, nuclear weapons, naval bases, and computers can shift the rules of a discourse. While material objects and events don't specifically determine any discourse, they play important roles. They have a great deal to say. Some things and events can't be ignored; others are evocative of particular ways of thinking. This is not technological determinism. Humans choose to make particular technologies and they choose to seek answers to particular scientific puzzles. These choices are complex, and they have ramifications of their own. Perhaps some choices make others easier, but they are never inevitable.

2. What is said or written (be it scientific theories, conventional wisdom, definitions, or other rhetoric) can have a direct effect on the discourse rules. This is especially true of any reflexive conversations, conversations about the terms of conversation, since they are about the metarules themselves. It is with words, especially texts now, that the weight of traditions is inscribed on the bodies of believers and nonbelievers alike, leading to physical actions and reactions that are particularly important elements in the postmodern war discourse.

3. Human actions, the actual physical deployment of bodies, can strongly impact the rules of the discourse. When people act with or on bodies, other humans take note. War is actually the best example of this (Scarry, 1985). As Hugh Gusterson points out in writing about the Gulf War, "In matters of contemporary war the human body can be both a blank page awaiting the violent inscriptions of power and, when its subjectivity escapes subjection, a source of embarrassment and resistance to that power" (1991, p. 45). War is all about what can be done to bodies. But protesting, buying, selling, loving, hating, creating, dreaming, and many other interesting and important statements are also only made by human bodies.

4. Human institutions have their own "say" in the discourse. In part this is because the human bodies institutions deploy, the things the humans in the institution say, and the material objects and powers of the institution, all contribute directly to the discourse. But it is also because the institution

itself, whether it's a church, a business, a university, or an army, organizes its own part of the discourse field.

Together, these categories not only explain in part how a discourse system is modified over time to maintain its hegemony but also how it is changed from without, perhaps even converted to a new regime of truth, by the insurrection of subjugated knowledges. How this takes place specifically may be the most important issue of all.

Constructing "Truth"

If this "regime of truth" is going to change, Michel Foucault argues that it must be through the "insurrection" and "reappearance" of subjugated knowledges which will have the effect of "changing . . . the production of truth" (1980, p. 133). What might this mean?

Let us consider two examples. First, in the instance of Vietnam, a number of factors came together to eventually reframe the Vietnam War as it was understood in American culture. From a noble attempt to preserve democracy it became an insensitive, if not evil, attack on a distant land. The rules that were used to justify and explain the war were influenced by all four of the forces just mentioned, so that eventually they came to discredit the war. The struggle over the meaning of the Vietnam War is not over. Vietnam has become a rhetorical figure in itself, and the conflict over exactly what it proves and implies may last for decades. The dynamics of such a struggle are very complicated. They range from the portrayal of Vietnam veterans in popular culture through a fascination with what contemporary presidential candidates did during the war to technical debates about strategic bombing.

A key part of this process is showing how certain metarules that are sacred in discourse in the United States (freedom, justice, and the American way) do not really lead to the rules that framed discussion of the Vietnam War. In turn, the metarules themselves can be questioned and even changed, slightly, in the very process of remaking the Vietnam discourse.

The impact of the new technology of television (including satellites) has been remarked upon by almost every observer of the Vietnam War. To this day military officers and conservative columnists advocate total news management (as was done with the Grenada invasion), because of the role of TV during Vietnam. It is hard to convince people of one thing when something quite different is on the news every night at 6:00 P.M.

On the level of what is said, a whole array of different versions of what was happening in Vietnam came to the fore, helped greatly by exposés of official lying and deception. Crucial voices in this insurrection of subjugated knowledges came from those who actually fought there (such as Vietnam

Veterans Against the War) and others such as AID (Agency for International Development) workers or exiled Vietnamese, who could speak directly from experience. The great value of direct experience is one of the key metarules of Western political discourse, especially in the United States, going back at least to the Protestant validation of the individual's experience of God.

When it comes to the application of bodily force as a statement of subjugated knowledges, the best examples are the hundreds of thousands of antiwar protesters. Their main statement was often just to go out into the street. Sometimes they got themselves put in jail as well, a symbolic and actual rejection of the dominant discourse. Other actions were aimed directly at the war system. Some were violent (bombing), some not (blocking trains, resisting the draft). The dominant discourse, of course, responded with bodily force of its own. What is important in these struggles is not who wins the body war but whose version of that war becomes dominant. The appeal is often rhetorical in word or image: a flower in a gun, a burning flag, a burning girl running down a street.

In the case of Vietnam, the uneasy acceptance of the liberal thesis that Vietnam was a mistake (as opposed to the radicals' version that it was business as usual or the conservatives' belief that it was right and necessary, just not carried through) is still being challenged from all sides.

Another, related discourse struggle, still far from over, is around the central subject of this book, the discourse of the contemporary military. On the level of artifacts, the reality of bombs that choose their own targets, computer displays in many colors, machines that talk, and all the other little working projects, there is powerful support for the military's discourse, even though the actual usefulness of the deployed weapons is problematic and the possibility of future weapons as promised is small. Artifacts are powerful arguments. They are evocative of meanings beyond any logical reasons because they appeal to unarticulated erotic, aesthetic, and mythopoetic desires.

In the arena of what is said, however, there is great importance in what is not there. Within the discipline of AI, with most of its rules and metarules taken directly from the discourse system of science, the continual failure of military AI proponents to produce what they say is possible is very important. Already in AI's short history failures to produce scientifically reproducible results have led to major shifts in the field, specifically in the case of a number of early research programs and the abandonment, by the military, of most natural language research in the late 1950s.

Within the military it seems only war, a great maker and unmaker of bodies and artifacts, can shift the discourse substantially. Vietnam made the Pentagon more careful politically but more extreme technologically.

The insurrection of subjugated knowledges still takes place on the physical level as well, of course. For example a woman named Katya went into the Vandenberg Air Force Base and smashed some computers for the Navstar system, part of the World War IV command-and-control network set up by the Pentagon with secret Black Budget funds. At her trial people were not allowed to mention "Nuremberg," or "International Law," or "First Strike." Even white roses were banned in the court because she chose the anti-Nazi resistance group of the same name as her symbol. She reached a few hundred people deeply with her act and thousands more read about her or saw 30 seconds about her on TV. The economy of punishments paid her back with three years in a cage to try and forge the link in all our heads: protest = prison. She has moved to subvert such an equation by earning early release and then graduating from Harvard Law School. Meanwhile, others took their hammers to military computers—such as Susan Rodriguez, found guilty of second-degree burglary and felony vandalism for using a sledgehammer on 55 computers at Physics International, which holds numerous military contracts (Gust, 1990; Martinez, 1990).

Dominant discourses are always shifting to take into account new events, and yet they are always trying to keep their metarules in force, such as "There will be war." But some of the most effective forms of metarule change can also be found in the continual struggle of subjugated knowledges. Such rule changes mark clearly the tautology that these power–knowledge systems that seem so out of control are the creation of human will whether we admit it or not. All of these forces in discourse can be deployed, either to benefit the continuation of the dominant discourse or in the insurrection of its subjugated knowledges.

It is important to notice that it is subjugated *knowledges,* not a single knowledge. No one person or theory will deconstruct this system and exceed the bounds of what we can consider possible. Foucault points out that the subjugated knowledges are many, their liberation and the breaking of the present power–knowledge system being a question of the politics of truth:

> It's not a matter of emancipating truth from every system of power (which would be a chimera, for truth is already power) but of detaching the power of truth from the forms of hegemony, social, economic and cultural, within which it operates at the present time. (1980, p. 133)

Redistributing power, for that is what Foucault is saying, is a matter of politics. And, everyone seems to agree, war is politics. But looking more closely it seems that something has happened to change this. The current system of postmodern war is "just the opposite of politics" Virilio and Lotringer claim (1983, p. 170). But leave it to Michel Foucault to turn Clausewitz totally inside out. For Foucault (1980), politics and power are "war continued by other means." He charts three implications from this thesis: First, that the role of political power . . . is perpetually to reinscribe this relation through a form of unspoken warfare; to reinscribe it in social institutions, in economic inequalities, in language, in the bodies themselves of each and every one of us. (pp. 90–91)

Second, that the political system is "the continuation of war." Third, that "the end result can only be . . . war." Two other French theorists, Gilles Deleuze and Félix Guattari, even go further and argue that the "war machine," as they call it, has "taken charge of . . . worldwide order" (1986, p. 133). If one accepts their claim (and I am ambivalent about it) that there is a "war machine" (I would say discourse) that is independent of the needs of states and other institutions, you'd have to agree that the current militarized world political system is a sign of its great success. There is certainly evidence that politics are now dominated by war.

How can this "politics as war" inversion be broken? One obvious way is through language, by directly intervening in the discourse system. Michael Shapiro argues for this: "To enlarge the realm of politics—to politicize more aspects of human relations—one must analyze language as a domain of political relations and thereby use it rather than be used by it" (1981, p. 233). It is clear that it is crucial to reveal the political and other purposes hidden in technical forms and claims of technological inevitability. Rhetorical and philosophical interventions, if they can find a hearing, certainly can be effective in changing a discourse based on the emotion of no emotion, the rhetoric of no rhetoric, and the politics of no politics, but it probably will not be enough.

We must go further with Foucault's reversal of Clausewitz. If the end of war is to come, and the end of the war = politics equation, politics as we have known it in a world of war will have to end as well. This points to the importance of the antipolitics, or the new politics, of the antiwar movements, the peace movements, and the other social movements of identities and issues. They are seeking to create a political discourse that is unconnected to war.

Yet it is also inevitably connected in some ways, especially by inversion or negation, for war is an integral a part of the current dominant Western discourse. That is why many have argued for a strategy of subversion. James Fallows, for example, suggests that what may be needed is what William James called for, an "intellectual and emotional equivalent of war," namely,

a way to talk out feelings about masculinity, about national pride, about the proper balance between force and moral example in international relations, without turning our military into a game board on which those feelings are secretly played out. (1982, p. 183)

Fallows is saying that the metarules around masculinity, nationalism, and the morality of force are negotiated and defined through military policy, instead of in the culture as a whole. It is a dangerous situation. He goes on to quote Larry Smith's call for "an analytical road map" that would differentiate disagreements "based on facts" and those "based on assumptions." But he warns, "it will never be neat and pretty." What he is calling for is a discourse analysis that keeps the metarules separate from instances everyone can agree upon. Perhaps it could lead to an unmediated discussion, but it seems unlikely. People have to share premises to reach agreements; otherwise they talk past each other. The real important issues aren't the facts, they are the assumptions. Different assumptions produce different sets of facts.

This analysis has tried to look at the assumptions behind the facts, at the ways some facts help shape assumptions, and how we might relate to all this complexity, and even change it. I agree with Smith when he admits, "It will never be neat and pretty."

Beyond Neat and Pretty

The technical–military discourse, which wields such power over our lives, is defended with incredible rhetorical venom as rational and logical and it is defended with professional, institutional, and other methods concretely. Yet, it is really not disengaged intelligence so much as disembodied/decapitated reason. The actual distinction is not between the rational and the emotional but between epistemologies. There is the dominant theory of knowledge that is alienated, unaware of its emotional dynamics and closed because it claims universality for one specific (hyper-logical-rational) way of knowing. And there are others that are not hegemonic, that are engaged, limited (by preunderstandings among other factors), but always open because of their explicit denial of universality in an ironic understanding of their own irony that goes beyond a sense of absurdity and beyond despair:

> What you have in the absurd is a passage clear through irony to a despair so awful that it becomes blackly comic instead of darkly tragic. It's very much a part of what I would call the postmodern atmosphere, where even the gestures, intellectual and artistic, of modernism seen insufficient to register the apparent absurdity of contemporary actuality. (Fussell, 1989b, p. 8)

This is perhaps because the absurdity is only apparent. Underneath the absurdity of Kurt Vonnegut's *Slaughter House Five* and Joseph Heller's *Catch-22* there is an awful sense of vertigo. They expose a void; the gap between our sensibilities and the reality of postmodern war. Transcendence will be found in closing that gulf, not in mining it for new wars.

In one of Foucault's last interviews, he used a metaphor for seeking truth that corresponds to the process that most activist peace groups use in their decision making. They call it "consensus" and it is based on traditions such as Quaker meetings and women's consciousness-raising groups. Foucault calls it "dialogic." The basic elements of both are the same: the refusal of the "enemy" dualism, the insistence on mutual rights, and the acceptance of limited "difficult" truths instead of allegiance to a "just cause" (Foucault, 1984, pp. 381–382). This begins by recognizing that there are no sources of unmediated truth. Both war and peace can be labeled natural, but what is important is that either is possible. In choosing, the value of living human bodies specifically and the living environment in general should be weighed against ideological and personal furies. And this judgment must be by everyone, not just the elite individuals and institutions that dominate contemporary societies.

It is a matter of survival. The Sarajevos of 1914, 1944, and 1995 should never be repeated. War is very strong. We must be stronger.

Notes

Introduction

1. The use of mass rape as a conscious policy by military units, as in the Balkans (Tax, 1993; Dobos, 1993) and in Iraq (Makiya, 1993) is hardly a new war strategy but it seems to be increasing. Feminist theorists who link this resurgence of one of the most ancient and horrific aspects of war to a general crisis of patriarchy and a particular crisis in war are very much on the mark (Enloe, 1993).

2. The good Harvard professor sees the era of modern wars between nations and ideologies as passé. Future conflicts, according to him, will be between the eight major "civilizations" of the world: Western, Confucian, Japanese, Islamic, Hindu, Slavic Orthodox, Latin American, and African. This bizarre, atavistic, and cartoonish theory has "hit a vein" Huntington admits, and it may well propel the old Cold Warrior to mass culture stardom (*Chronicle of Higher Education*, 1994, p. A10).

3. The information in this section comes, appropriately enough, from the Internet—specifically from a series of postings forwarded by the Austin (TX) Comité de Solidaridad con Chiapas y México, including notes by Joseph Moore (y@igc.apc.org), Harry M. Cleaver (hmcleave@mundo.eco.utexas.edu), and translations from Zapatista bulletins. For more information contact: (Chiapas96@mundo.eco.utexas.edu).

4. In this book I hope to explain contemporary war and show its relationship to postmodernity in general. To this end there will be several extensive attempts to define and explain both the specific elements of postmodern war and the general idea of the postmodern, starting in Chapter 1.

5. "One can no longer distinguish between technology on the one hand and theory, science and rationality on the other. The term techno-science has to be accepted" (Derrida, 1984, p. 12). In the case of military computing the term "technoscience" seems quite applicable and I will be using it (without the hyphen) unless I am pointing to something specifically within science or engineered technology, but not both.

6. *He* in this case but not in feminist science fiction, where the cyborg is often a *she*.

Chapter One

1. The postmodern soldier would say, "Victory in war can only be based on the latest information." Is this the same information?

2. Reuters, "Balkan Talks Strained Allies' Ties," *International Herald Tribune*, November 25–26, 1995, p. A2.

3. Calling the United States the "only" superpower in the post-Cold War era is dangerous. Actually, in terms of either economic (constructive/coercive) power or purely military (destructive/coercive) power or both, there are now dozens of superpowers in the world, including a number of multinational companies. The United States has a special position, certainly, but is hardly omnipotent, not even in culture, where it is strongest. Still, it is not insignificant that U.S.-based companies are reporting record profits and foreign investments ($33 billion in the first half of 1995, up 27 percent over the previous record year, 1993) as the Pax Americana gets on its way. See Allen R. Myerson, "U.S. Firms: The Worldly Shoppers—International Investment Totals Reach Record Levels," *International Herald Tribune*, November 25–26, 1995, p. B1. Worldwide international investment for the first half of 1995 was also a new record, $226 billion.

4. On the other hand, some of the theories are far from useful. Baudrillard's (1991) claim that the Gulf War was merely simulated isn't helpful even as overblown postmodern rhetoric, as Norris (1994) and Nideffer (1993) among others have made clear. The conceit in Manuel De Landa's *War in the Age of Intelligent Machines* (1991) that AI robots are taking over is both technically wrong (AI is very distant, if not impossible) and hopelessly deterministic.

5. This term was initially used by Fredric Jameson (1984) when he labeled Vietnam the first postmodern war in his article "Postmodernism, or the Cultural Logic of Late Capitalism." Ann Markusen and Joel Yudken (1992) use the same term in their book, *Dismantling the Cold War Economy*, although their conception of modern war is ahistorically limited to the twentieth century. There is a school of thought (Nadel, 1995; Kellner, 1997) that sees the postmodern break as being precipitated by the Vietnam War but coming after it. In many ways it takes a similar approach to my own, and I am in broad agreement with many of its conclusions.

6. Douglas Waller, "Onward Cyber Soldiers," *Time*, August 21, 1995, pp. 39–46; Newt Gingrich, "Information Warfare: Definition, Doctrine, and Direction," address to the National Defense University, Washington D.C., May 3, 1994; Jackson Browne, "Information Wars," *Looking East*, Elektra Records, 1996; Tom Clancy, *Debt of Honor*, 1994; Ann Shoben, ed., *RAND Research Review: Information War and Cyberspace Security*, 19, no. 2, Fall 1995. The Clancy novel is cited most often of all of these in the many military articles on information war.

7. Some infowar advocates see it as being limited to electronic and electromagnetic weapons; others argue that it is the role of information that is important. Different definitions of C^4I^2 war, infowar, netwar, and cyberwar proliferate. The distinctions aren't important. What is crucial is to see the continuity of these types of war with the long history of low-intensity (LICs) conflicts, on the one hand, and with the existing system of postmodern war, on the other.

8. OOTW was first included in the U.S. Army's FM (Field Manual) 100-5

military master plan in 1993. In war's crisis war has become as flighty theoretically as Parisian fashion. Robert Bunker's clear critique of OOTW even reaches the point of complaining that operations other than war aren't categorized under the general category of war (1995, p. 40). There is a simmering debate over OOTW and LIC and other acronyms, but no matter how you slice the pie, or what you call it, there are only so many military missions up for grabs.

9. The main historical framework that most infowar theorists use is that of Alvin and Heidi Toffler. For them, infowar typifies the Third Wave of human culture, the information wave as opposed to the industrial and agricultural epochs that preceded it. In their 1993 book *War and Anti-War* they counterpoise brute force and brain force and hail the spread of cyberwar doctrines, nonlethal technologies, and electronic democracy initiatives. Facile as much of their analysis seems, the general sweep of it is unpleasantly compelling. Even if one accepts that a different schema is better, such as Bunker's four epochs based on energy use (1994), it does seem that the Tofflers' claim that humanity is undergoing a technologically based cultural revolution is all too real.

10. Neil Munro, "The Pentagon's New Nightmare: An Electronic Pearl Harbor," *Washington Post*, July 16, 1995.

11. Described in graphic detail in Mark Thompson, "If War Comes Home," *Time*, August 21, 1995, pp. 44–47.

12. The U.S. Army has Force XXI; the U.S. Navy has its program "Forward . . . From the Sea"; the U.S. Air Force has "Global Reach, Global Power"; and the U.S. Marine Corps has "Operational Maneuver . . . From the Sea." For details see: Col. Richard Szafranski, "A Theory of Information Warfare," *Airpower Journal 4*, no. 1, Spring 1995; Col. G. I. Wilson and Maj. Frank Bunkers, "Uncorking the Information Genie," *Marine Corps Gazette*, October 1995, and other articles to be found at Dr. Ivan Goldberg's web site called the Institute for the Advanced Study of Information Warfare: <www.psycom.net/iwar1.html>. The Air Force has taken this furthest with the implementation of the Copernicus Project in 1990. It sets up an infowar architecture for the whole Navy, as is described in "Copernicus Forward 1995," Navy Public Affairs Library.

13. Richard Sisk, "In Age of High-Tech Cybergrunt, Old Soldiers Never Lie," *Oregonian*, August 28, 1995, p A7; Chris Black, "US Options Seen Fewer as Military Avoids Risk," *Washington Post*, July 26, 1994, p. A3.

14. This is a dangerous continuation of the LIC theory that argues that domestic enemies were responsible for the Vietnam War defeat, as is detailed in Chapter 9. Charles Swett (1995), for example, who works for the assistant secretary of defense for special operations and low-intensity conflict, has written a "strategic assessment" of the Internet that focuses mainly on domestic political protesters, although the use of the Internet by the Zapatistas is a major case study, as it is with all of the cyberwar theorists. See also David Corn, "Pentagon Trolls the Net," *The Nation*, March 4, 1996. For current information on this debate access the Federation of Atomic Scientists' (FAS) excellent site: http:/www.fas.org/pub/gen/fas/cp.

15. See http://www.is.in-berlin.de:80/pit/WEBSTOP/CAE/cae_ed1.txt for this and more recent manifestos.

16. Chris Hedges, "Land Mine in Bosnia Kills American Soldier," *Oregonian*, February 4, 1996, pp. A1, A3. On March 15, 1996, the United States announced it might join international attempts to ban land mines (Public Radio: "All Things

Considered," March 17, 1996). The United States has also changed its longstanding deployment policy, finally recognizing that experience/information is crucial to survival. Soldiers assigned to the Bosnian mission will be there for the duration ("Deployments Will Be Longer," *Oregonian*, March 6, 1996, p. 7).

17. Remy Allouche and Nathalie Bichat, "Multimedia Products Respond to Evolving Crisis Management, Peacekeeping Needs," official Armed Forces Communications and Electronics Association publication, reprinted from the March 1995 issue of *SIGNAL Magazine*. There are a number of such network systems for international missions, including Tempo, Pegas, Takom, and Alcatel. Many stress tie-ins to civilian technology such as the Internet in order to influence local events and also to allow soldiers to e-mail to each other and home.

18. Quotes and information from the meeting of the Czech Armed Forces Communications and Electronics Association, "Ceská Poböka," in Brno, Czech Republic, October 22–23, 1995.

19. In many instances, according to Yugoslav peace activists I've interviewed extensively (and who wish to remain unnamed) and Masha Gessen (1995), effective communication among peace activists within the former Yugoslavia has only been possible because of the activist-created computer networks, which continue operating, with difficulties, into 1997.

20. On January 5, 1996, Israeli intelligence killed Yehiya Ayyash, one of the leading bomb makers of Hamas. For the next two months Hamas staged a series of suicide bomb attacks. Sheikh Jamil Hamami, a senior Hamas official, said that "Hamas militants then had no choice but to avenge Ayyash's killing." See Scheherezade Faramarzi, "Arafat orders Hamas to Give Up Arms," *Oregonian*, February 29, 1996, p. A3. Vengeance still trumps peace in this conflict, as the Israeli responses demonstrated as well.

21. Shawn Pogatchnik, "Britain Posts More Soldiers Along Northern Ireland Border," *Oregonian*, February 11, 1996, p. A4.

22. Julia Preston, "Zapatista Rebel Leaders Agree to Sign Agreement with Mexican Government," *Oregonian*, February 11, 1996, p. A4.

23. Alvin Bernstein in a speech at the University of Maryland, College Park, July 28, 1990; emphasis added.

24. This paragraph is based on Teresa O'Connell's 1990 unpublished manuscript, "From Scorched Earth to Scourged Earth: Drugs and Destabilization in Guatemala."

25. Fort Monroe, Virginia, August 1, 1986. Only Vol. 1 is available for analysis, however, as Vol. 2 is "Top Secret." All subsequent quotations in this section are from Vol. 1 of this review unless otherwise noted; their page numbers will be given in parentheses.

Chapter Two

1. Throughout this chapter I will use various names for the 1990–1991 conflict between Iraq and the alliance of Kuwait, the United States, Egypt, Saudi Arabia, Great Britain, France, and many others. The war has no official or accepted name as yet, so it seems appropriate and interesting to try and use those I've heard. My personal favorite is "The War to Restore the Rightful Dictator of Kuwait."

2. Film from the Apaches taken during the U.S. invasion of Panama is still being withheld, even from Congress. Congressman Charles Rangel (Dem., N.Y.) suspects that the film shows Panamanian civilians being killed (Rangel, 1991).

3. Originally, the attack on Iraq and occupied Kuwait was to be called Desert Sword, but it was decided to portray the war as more of a natural force. Usually, all military operations are given random computer-generated names, but not in this case, which at least shows some understanding of the limits of computers. Perhaps Secretary of Defense Richard Cheney made the change. He renamed the invasion of Panama Operation Just Cause after the computer gave it the designation "Blue Spoon" (Waller et al., 1990).

4. Estimates vary widely. Iraqi military casualties are put by many different sources at 100,000–200,000, most of whom were untrained conscripts. Direct civilian deaths in Iraq from Allied bombing are put at less than 5,000. Indirect deaths were estimated at 70,000 after one year by the U.S. Census Bureau, with 100,000 more, mainly children, two years after the conflict, according to a Harvard University study. Deaths in the consequent civil war have been put at 20,000–35,000, mainly civilians (Weiner, 1992).

5. In all likelihood this was a decision made out of local enthusiasm, although it could represent a military protest to the precipitous cease-fire. In retaliation for "two rocket-propelled grenades and a single round from a T72 tank" fired at a U.S. patrol (probably by mistake) and causing no casualties, a whole division was savaged (Gordon and Trainor, 1995, p. 429). Under the protection of the truce that the United States called, the Hammurabi Division of the Republican Guard was traveling up Route 8 along the Euphrates River to Baghdad. While it was crossing the Haw al Hammer swamp on an elevated highway, Gen. Maj. Barry McCaffrey's 24th Mechanized Infantry Division attacked it without warning two hours after the firing incident. Using helicopters, artillery, and armored vehicles, they destroyed hundreds of vehicles, captured 3,000 prisoners, and probably killed thousands of others, all in a matter of a few hours at the cost of one soldier wounded, one tank lost, and one Bradley armored vehicle damaged. Despite standing orders to the contrary, the Iraqis were not warned or asked to surrender. The Pentagon kept the attack from the press at the time by consciously mislabeling it as a series of small actions (Sloyan, 1991; Gordon and Trainor, 1995, p. 429).

6. The main contractors for these systems include IBM, GTE, Honeywell, Computer Science, Unisys, and Boeing. Many other companies are involved as subcontractors, including Lockheed, a key satellite manufacturer, Cisco Systems, Inc., and even Apple Computers, which has installed computers in the National Military Command Center and on helicopters. There were also numerous small specialty companies, such as Rugged Digital Systems and Grid Systems, with computers in the Gulf (*Business Week* Staff, 1991c, p. 43; Clark, 1991).

Chapter Three

1. There is a difference between the Strategic Defense Initiative (SDI) and Star Wars: SDI was a package of specific programs; Star Wars is the ongoing militarization of space. SDI was a mythological "peace shield" with a specific budget, now ended, although many of the projects live on in other bureaucratic homes. Some leading

Star Warriors, such as the director of SDI research at Lawrence Livermore in the mid-1980s, felt the U.S. military was doing fine in space before SDI. They objected to Reagan's impossible system, not realizing the tremendous political victory his proposal would become [Gen. C. Neil Beer, USAF Ret., director of SDI research at Lawrence Livermore Laboratories, interview (May 19, 1986) and personal correspondence (June 3, 1986)].

2. At the 1987 AAAI (American Association for Artificial Intelligence) conference SDI and SCP were the subjects of much official and unofficial debate. Over 7,000 U.S. scientists pledged not to work on SDI, for example, including many computer scientists. For technical views on this debate see the Eastport Study Group Report (1985); Ornstein et al. (1984); Parnas (1985); B. Smith (1985); and Nelson and Redell (1986).

3. This list is compiled from official DoD Strategic Computing Program documents and Morton (1988, pp. 39–43; quote from p. 41).

4. Douglas Waller, "Onward Cybersoldiers," *Time*, August 21, 1995, pp. 38–44.

5. From *Defense Computing* magazine promotional material, 1988, citing a 1987 study by the Electronics Industries Association and from Robert J. Bunker, "Digital Battlefield Conference," *Military Review*, May–June, 1995, pp. 3–4. However, *New York Times* reporter Jeff Gerth (1989) put annual Pentagon spending on computers, software, and related services at $9 billion. It wasn't clear if either of these figures include computer expenses of the CIA, NSA, NASA, and DEA, of the DOE's weapons manufacturing, of ARPA and other research, of the microchips in weapons, or other hidden expenditures directly in the Black Budget. In any case the amounts spent are huge.

6. Corporations with SCP contracts in the 1980s were Martin Marietta, Ford Motor Co., General Dynamics, ADS (Advanced Decision Systems), Vitalink, Computer Systems Management, Texas Instruments, Computer Corp. of America, Lockheed, McDonnell Douglas, MITRE, Intellicorp, Bolt Beranek & Newman, Cognitive Systems, LOGICON, Titan Systems, The BDM Corp., Science Applications International, Symbolics, Mark Resources, MRJ Inc., The Analytic Science Corporation, Northrop, Hughes, AVCO Research Laboratories, SRI International, General Electric, Fairchild, Dragon Systems, Systems Development Corp., Teknowledge, Thinking Machines, Incremental Systems, Kestrel Institute, Trusted Systems, Honeywell, Rockwell International, Westinghouse Electric, Optivision, Sperry Computer Systems, Sparta Inc., and Cyberoptics Corp.

7. A partial listing of universities with researchers accepting SCP contracts in the 1980s shows broad participation: Carnegie–Mellon University, University of Maryland, University of Massachusetts, University of Southern California, Massachusetts Institute of Technology, University of Rochester, University of Pennsylvania, Columbia University, New York University, Ohio State University, Stanford University, Yale University, University of Minnesota, Georgia Institute of Technology, University of California at San Diego, University of Southern California–Information Science Institute, Brown University, University of California at Los Angeles, University of Cincinnati, University of Houston, Northwestern University, City University of New York, Columbia University, University of California at Berkeley, Princeton University, and the California Institute of Technology.

8. An expert system is an AI program that simulates the information and logical

processes of some expert in some limited and carefully defined field. Currently there are civilian expert systems that analyze geologic data to predict if oil might be found, analyze X-rays and other medical data for diagnosis, and repair engines, among many others tasks.

9. Another reason for the failure of the Libyan raid (unless you count killing Gadhafi's 4-year-old daughter as a success) was the low radar "significance" (meaning reflectivity) of Tripoli in general. Despite laser targeting that involved shining laser beams on the targets, and target confirmation through radar and infrared TV pattern matching, only a few of the targets were hit while stray bombs struck the French embassy and many private houses (Editors of Time–Life, 1988, pp. 61–62).

10. The Time–Life book *Understanding Computers: The Military Frontier* has some excellent illustrations of several of these games, including RSAS (see Editors of Time–Life, 1988).

11. From Meyer (1988, p. 103). Others argue that the Aegis, while "a giant step up the ladder of automation in combat systems" because of its self-testing, automated detecting, and total hands-off capabilities, is still "not an application of expert systems technology" in a full sense (Keen, 1988, p. 98). Just exactly what counts as being AI work is a matter of debate and continual redefinition. The Aegis can be called an artificial intelligence for several reasons. First, it fits most dictionary definitions of AI (Beardon, 1989, p. 12; A. Chandor et al., 1981, p. 38). Second, the military clearly thinks of it as an AI system, and that is a key reason it has had the effect it has. And, third, although the details of its programming are highly classified, it seems to me that any system that coordinates so many complex subsystems, that integrates so much different data, that controls so many deadly weapons, and that has the independent ability to find a target, choose the target, and destroy that target, as the Aegis does, is an AI. For a full discussion of the Aegis system in action see C. H. Gray (1997).

12. When categorical statements are made about what the Aegis noted or didn't, they are based on the Navy's replaying of the *Vincennes'* Aegis tapes in a mock combat information center at the Aegis Combat System Center at Wallops Island, Virginia. I am assuming in this case that there were no recording errors and that the information was not tampered with, although both are certainly technically possible.

Chapter Four

1. See the work of Marvin Minsky for the *frame* metaphor and a sophisticated mix of many of the other current favorites in his "society of mind" idea (1985). Schank and Abelson (1977) are the main proponents of *scripts*. Simon (1980) and Newell (1986), alone and together (1963), have done the most to spread the simplistic "rational man" model.

2. As is his fashion, Latour not only rejects the term *postmodernism* in this essay but argues that there was never any modern world at all. He labels the current era *premodernism* or *nonmodernism* (see especially p. 17).

3. Other efforts at reconceptualizing science from feminist and related perspectives can be found in many different grass-roots groups. Especially numerous are groups that

advocate the revaluation of nature, such as eco-feminists, Greens, anarchists, deep ecologists, and neopagans (Epstein, 1986). For a Marxist-feminist argument for adding "heart" to the "head" and "hand" of modern science see Rose (1983).

Chapter Five

1. Ferrill traces modern war back as far as Alexander the Great, and even argues that Alexander's army could have beaten Wellington at Waterloo. While a bit far-fetched, this contention does make his point that in terms of scale and weapons battle did not change that much, especially in style, in the c. 2,150 years between the battles of Issus (333 B.C.E.) and Waterloo (1815 C.E.). I agree with most historians who argue that the technological changes (in weapons, transportation, and communications), sociological changes (in class structure, bureaucratic infrastructure, and army composition), and philosophical or methodological changes (in leadership, logistics, and the art and science of battle) in war around the sixteenth century mark a new type of war, modern war. What Ferrill's argument rightly emphasizes is that between ancient and modern war there is a great deal of continuity.

Chapter Six

1. Some people would use "modernization" here in place of modernism. I make the term an "ism" in order to emphasize how much the modern point of view is a matter of belief and is not just a given. Modernism is an ideology constructed around the central tropes, rhetorical and ontological, of individualism, science, technology, progress, productivity, art, nationalism, and democracy, among others.

2. The ecological historian Alfred Crosby coined this term to refer to the Americas, Australia, New Zealand, and South Africa, where European people, flora, and fauna have come to dominate the indigenous species (1986).

3. Of course the story is much more complicated than that. As Alfred Crosby shows in his book, *Ecological Imperialism: The Biological Expansion of Europe, 900–1900* (1986), disease played a major role in weakening indigenous people. There were other important factors as well. Yet, when all is considered, the decisive element seems to have been the success of modern war.

4. This explains in part why Hall and Wolf downplay the role of technoscience in the developments of war during this period; they neglect the importance of smiths and other craftsmen in the founding of science. Both Edgar Zilsel (1942) and Mircea Eliade (1962) have shown that such groups were a crucial and often neglected part of the creation of modern science.

5. Comte de Guilbert, *Défense du système de guerre moderne*, 1789, cited in Macksey (1986).

6. See Catton (1981). He argues that the Civil War was the first modern war. It did mark the decline of cavalry as a first-line military arm, the first ship sunk by a submarine, the first ironclad duel, the use of railroads for logistics, and the extensive use of balloons, rifled and repeating small arms, and other innovations. But is this

enough to make it the first modern war? Modern war seems to have been developing consistently since the end of the feudal era, as is argued in the previous chapter. The Civil War is in complete continuity with the Napoleonic campaigns, where artillery was often decisive, and even further back to the total war waged by the French Republic, reinventor of conscription in modern times.

7. The studies involved testing whisky, preserving paint on knapsacks, the galvanic action of zinc and iron, exploring Yellowstone, and meteorology (Kranzberg, 1969, pp. 125–126).

8. From Ellis (1973, pp. 81–82, quoting V. A. Majendie). Machine guns, only possible because of mass production techniques, were also very important in the suppression of mass labor protests, as Ellis points out.

Chapter Seven

1. Box 8, records of the 20th Air Force, File: Operations Analysis Studies No. 8, *Targets and Target Damage*, Vol. 1. Operations Analysis Division, 20th Air Force. Over a quarter of a million dead were predicted in Tokyo after four attacks.

2. Box 18, Records of the 20th Air Force, File 373.2, Report of Operations, General. "Interoffice Memorandum," October 12, 1944.

3. Box 17, 20th Air Force, File: Telecons concerning bombing of 20th AF, Kyushu Airfield, 373.11, April 25, 1945.

4. Box 16, 20th Air Force, File 373, Employment of Aviation. Report of March 22, 1945, 20th Air Force Planning Factors, by Lt. Col. R. S. McNamara, operating assistant, statistical control, 20th AF.

5. Box 16, 20th Air Force. File 373. Employment of Aviation. "Statistical Summary of 20th AF Combat Crews," March 12, 1945.

6. Box 18, 20th Air Force, File 373.2, Report of Operations—General. "Comparison of Actual vs. Planned Strategic Bombing Accomplishments for Month of January." From Lt. Col. R. S. McNamara to Brig. Gen. L. Norstad, February 1945.

7. The war also played a key role in introducing cybernetic thinking to other disciplines. Conrad H. Waddington, the well-known geneticist, worked for the Royal Air Force Operations Research Section on defense calculations for combating U-boats. After the war he brought cybernetic paradigms into biology (Haraway, 1981–1982). Paul Edwards (1986a) convincingly traces the founding of the field of cognitive psychology to OR on antiaircraft gun control.

8. Just exactly what makes a machine the first computer is debatable and probably not that important. The idea of computing is an old one. The first "computers" so named were, of course, humans who did calculations. Even into the 1940s the large number of engineers and mathematicians, many women, who did calculations on mechanical and electromechanical machines for aeronautic and other companies were called "computers." The simple electrical and/or mechanical machines that did ballistic and other calculations, often using analog methods, for weapons and weapon platforms were also often called "computers." In the frenzy of World War II a number of machines were built that performed functions previously accomplished by human computers or those impossible for humans to do. They had various elements of what

we now recognize as computers, and many of them have some claim to being the first computer of some type or another.

9. Box 16, 20th Air Force Records, File: 373, Employment of Aviation, Memo to Brig. Gen. L. Norstad from Col. E. E. Aldrin, Army Air Corps, October 19, 1944.

10. Box 8, 20th Air Force Records, File 319.1. Daily Activity Report AC/AS OC&R. Letter from Brig. Gen. William McKee to Gen. H. H. Arnold, July 14, 1945, and letter from Maj. Gen. Donald Wilson to Gen. H. H. Arnold of June 22, 1945.

11. National Archives Record Group U-234, Box 8.

12. Merrill Flood, a RAND mathematician, claims to have coined the term "software" in 1946 to "distinguish costs not directly attributable to military hardware." When General Eisenhower saw the term he found it objectionable (Ceruzzi, 1989a, p. 249).

13. Two years later, RAND and 18 aircraft companies founded SHARE, a user group for the IBM 704. Led by Paul Armer, head of RAND's Numerical Analysis Department, it was the first computer user group (Ceruzzi, 1989a, p. 47). RAND also was interested in artificial intelligence from the beginning. Two AI "fathers," Allen Newell and Herbert Simon, wrote what might well be the first official military report on AI, RAND Research memorandum RM-2506, dated December 28, 1959, called "The Simulation of Human Thought" (p. 240).

Chapter Eight

1. For example, Gen. Hap Arnold in 1946. Many other military leaders have said that war must now end. See the quotes from Gen. Dwight D. Eisenhower, Gen. Omar Bradley, Adm. Charles Brown, and Soviet Maj. Gen. N. A. Talenski in the Center for Defense Information booklet, *Quotes: Nuclear War* (1983).

2. Kull, 1988, p. 240. At a conference I attended once, three different high U.S. officials—an adviser, a general, and an ambassador—said exactly the same thing. Unfortunately, I promised not to quote them.

3. For an eyewitness impression of the bombing of Hanoi see Salisbury (1967, pp. 188–198). He cites Hanson Baldwin of *The New York Times* for the number of U.S. planes lost. For a detailed account of the surface rivalry that warped the bombing campaign, and for an analysis of why opening up the restricted target areas in North Vietnam would not have changed the outcome of the war, see Baritz (1985, pp. 246–252). For a sample of the belief that more bombing would have won the war see Moorer (1987): Adm. Thomas H. Moorer (ret.), who had been the Chairman of the Joint Chiefs during the latter part of the war, claimed that if the military would have been given a free hand "we could have polished those clowns off in six months."

Chapter Nine

1. Throughout the JLIC the importance of definitions is stressed again and again. For example, to begin Chapter 7, "Peacekeeping," there is an extended discussion of just exactly what peacekeeping is:

The United States Army defines peacekeeping as military operations conducted in support of diplomatic efforts to achieve, maintain, or restore peace in areas of potential or actual conflict. However, no single definition is accepted by all services and agencies. (p. 7-1)

Then follows much tortuous explaining of the differences between peacekeeping and peacemaking and the admission that while some feel the military is not the appropriate way to try keep the peace, "the United States uses military forces as the primary element in such operations" (p. 7-1).

2. All three epigrams are quoted from Dillon (1988, pp. 15, 23).

Chapter Ten

1. *Combat, Hogan's Heroes,* and *McHale's Navy* were TV shows in the 1960s about World War II. *The Green Berets* was a best selling Robin Moore novel, a hit record, and a John Wayne movie. *Deerhunter, Platoon, Apocalypse Now, Coming Home,* and *Fields of Stone* are movies about the Vietnam War. *Tour of Duty* was a late-1980s TV series about a platoon in Vietnam during 1968–1969. *M*A*S*H* was a successful book, movie, and TV comedy about a Mobile Army Service Hospital (MASH) during the Korean War. *Good Morning Vietnam* is a 1980s movie about a wild disc jockey in Vietnam. *China Beach* was a 1980s and early 1990s dramatic TV series centered around nurses in a hospital in Vietnam, also during the crucial 1968/9 period. On the World War I poets see Paul Fussell's remarkable work *The Great War and Modern Memory* (1975). The definitive World War II novel is perhaps Joseph Heller's *Catch-22,* though Norman Mailer's *The Naked and the Dead* and Kurt Vonnegut's *Slaughter House Five* are equally extraordinary in their own ways.

2. Information from the National Coalition on Television Violence and by watching TV with my toddler son from 1985 to 1989.

3. See Smoler (1989) for a detailed account of how U.S. combat veterans helped reveal that the claims of S. L. A. Marshall's *Men Against Fire* (1947) were not true.

4. "Stanzas found on a leaf of an International Brigader's notebook," in Stallworthy (1984, p. 234).

5. Quoted at the head of Chapter 8, "Counterterrorist Commandos: The New Samurai," in Neil Livingstone's *The Cult of Counterterrorism.* Livingstone (1990) goes on to explain on p. 293, that "Some have called them the 'modern Ninja,' others the 'new Samurai.' They are a special breed of men, living testimony to the difference between mere soldiers and warriors."

Chapter Eleven

1. The following information, and the quotations, are taken from a copy of the U.S. Army's *AirLand Battle 2000* plan, dated August 1982. While the doctrine is in

place and being used, as in the Gulf War of 1991, many of the technologies on the DoD wish list are still under development.

2. Those figures are from Gen. Lawrence Skantze's speech "AF Science & Technology—The Legacy of Forecast II," given at Aerospace '87, the AIAA convention, Crystal City, Va., 1987.

3. Epigrams from *Military Space* Staff, 1988e, p. 1. Some European nations have proposed a unified European military space force, whereas others have called for demilitarizing space. Already there is a high level of European civilian space cooperation and various cooperative military projects such as the Italian, French, and Spanish joint military satellite (milsat) projects. France is especially committed to maintaining its great power status by having a strong military space presence. Israel and China also have milsat programs.

Chapter Twelve

1. To this day the story persists among radical computer scientists that Turing was murdered by British (or American) secret agents, although there is no direct evidence for this. The supposed method of death, a poisoned apple, is evocative though. Andrew Hodges has written a moving biography, *Alan Turing: The Enigma* (1983), about this brilliant and tragic scientist.

2. My thinking on von Neumann and Wiener, and on much else, has been strongly influenced by Eglash (1989).

3. All quotations by Theodore Taylor here are from public discussions at the University of California at Santa Cruz, April 28 and 29, 1986.

4. The peace army was endorsed by the National Mobilization for Survival and numerous other activist groups in the United States but never really got off the ground.

5. This is based on my 20 years as a participant-observer in various peace organizations and campaigns. It is true that generally the more formalized and bureaucratic an organization gets the fewer women are in it, but this isn't always the case, witness the Nuclear Freeze Campaign. For a moving and extended discussion of the warrior metaphor as it applies to women, see the issue of *Woman of Power* magazine dedicated to the "Woman as Warrior," 3, Winter/Spring, 1986.

Chapter Thirteen

1. That conventional war will die out and only LICs continue on infinitely as Van Creveld (1991) argues seems very unlikely. He seems to believe it only because he thinks war is inevitable, but apocalypse (like peace) is unthinkable.

Glossary

AAAI American Association for Artificial Intelligence
ABC American Broadcasting Company
ABCCCIII Airborne Battlefield Command and Control Center (III)
Ada The official computer language of the U.S. military. Named for Ada Lovelace, the first computer programmer.
ADS Advanced Decision Systems
AES Applications of Expert Systems
AFB Air Force Base
AGLT Automatic Gun-Laying Turret
AI Artificial Intelligence
AID Agency for International Development
AIDS Advanced Immune Deficiency Syndrome
ALB AirLand Battle
AR&DA Army Research & Development and Acquisitions (magazine)
ARPA Advanced Research Projects Agency. A Department of Defense agency that funds applied and other military research. Known as ARPA from its founding in 1958 to 1972, DARPA from 1972 to 1992, and ARPA again now.
Asats Anti-satellite satellites and other systems.
ASW Anti-Submarine Warfare or Association of Scientific Workers, a twentieth-century leftist scientist organization in Great Britain
ATF Bureau of Alcohol, Tobacco, and Firearms
AT&T American Telegraph and Telephone Company
BINAC Binary Integrator and Computer.
Bit Binary digit. A 0 or 1. The basic unit of computerized information.
Bips Bits per second. A measure of computing speed.
BMOS Black Moving Objects. Slang used by U.S. troops in the Persian Gulf War to describe Saudi Arabian women.
BUIC Back-Up Interceptor Control. A part of the SAGE system.
C^2 Command and Control

C^3I Command, Control, Communications, and Intelligence

C^4I^2 Command, Control, Communications, Computers, Intelligence, and Interoperability

CBS Columbia Broadcasting System

CBN Chemical, Biological, and Nuclear. Usually referring to weapons.

CDI Center for Defense Information

CIA Central Intelligence Agency

CNN Cable News Network

COA Committee of Operations Analysis

Comsat Communication satellite

CPSR Computer Professionals for Social Responsibility

Cyborg Cybernetic organism. Natural–artificial integrated system.

DARPA Defense Advanced Research Projects Agency. A Department of Defense agency that funds applied and other military research. Known as ARPA from its founding in 1958 to 1972, DARPA from 1972 to 1992, and ARPA again now.

DC^3I Distributed Command, Control, Communication and Intelligence

DEA Drug Enforcement Agency

DIVAD Division Air Defense System

DOD Department of Defense

ESP Extrasensory Perception

ENIAC Electronic Numerical Integrator and Computer

FAE Fuel–Air Explosives

FAS Federation of Atomic Scientists

Fatsats Larger than normal satellites

FBI Federal Bureau of Investigation

Flip-flop A circuit holding one bit of information.

FMLN Frente Militar de Liberación National

Gflips Giga-flips. One billion flip-flops.

GPALS Global Protection Against Limited Strikes

HQ Headquarters

HUMMRO Human Resources Research Office of the U.S. Army

IBM International Business Machines, Incorporated

ICBM Intercontinental Ballistic Missile

ID Identification

IDA Institute for Defense Analysis. The think tank of the Joint Chiefs of the U.S. military.

IEEE Institute of Electrical and Electronic Engineers.

IFF Interrogation Friend or Foe

Info War Information War

INS Immigration and Naturalization Service

IRS Internal Revenue Service

ISA Institute for Advanced Study

JLIC Joint Low-Intensity Conflict Project, U.S. Army Training and Doctrine Command

JOHNNIAC A RAND computer built in 1953 and named in honor of John von Neumann

JSTARS Joint Surveillance and Target Attack Radar System
LIC Low-Intensity Conflict
Lightsats Small satellites
LISP LISt Processing. A computer language mainly used in AI research.
LSD Lysergic acid diethylamide. A powerful hallucinogen.
LRRPs Long Range Reconnaissance Platoon members, pronounced "Lurps."
MAD Mutual Assured Destruction or Magnetic Anomaly Detection system
MADDIDA Magnetic Drum Digital Differential Analyzer
MACIPS Military Airlift Command Information Processing System
MANPRINT Manpower and Personnel Integration
MAPP Modern Age Planning Program
MASH Mobile Army Service Hospital
Milsat Military satellite
MIT Massachusetts Institute of Technology
Mflips Milliflips. Thousands of flip-flops per second.
MOSAIC Ministry of Supply Automatic Integrator and Computer
NAREC National Cash Register Electronic Calculator
NASA National Aeronautics and Space Administration
NATO North Atlantic Treaty Organization
NBC National Broadcasting Company or Nuclear Biological Chemical, as in weapons
NCI National Counter-narcotics Institute
NFL National Football League
NSA National Security Agency
OA Operations Analysis
ODS Net Operation Desert Storm network
ONR Office of Navel Research. Agency in the U.S. Navy that funds much applied and other research.
OOTW Operations Other Than War
OR Operations, or Operational, Research
ORSA Operations Research Society of America
OSS Office of Strategic Services. The World War II precursor to the CIA.
Project RAND See RAND.
Project SCOOP See SCOOP
PSYOPS Psychological operations
RAF Royal Air Force of the United Kingdom
RAND Research and Development. Originally Project RAND. A civilian agency to perform OR and other research on military problems. Named by Arthur Raymond. Headquartered in Santa Monica, California. Established by Army Air Force Contract on March 1, 1946.
Recon Reconnaissance
R&D Research and Development
RD&A Research, Development, and Acquisitions
RMA Revolution in Military Affairs
RPV Remotely Piloted Vehicle
RSAS RAND Strategy Assessment System

SA Systems Analysis

SAC Strategic Air Command

SAFE The CIA's master computer program. Perhaps descended from the old RAND SAFE (Strategy and Force Evaluation) game.

SAGE Semi-Automatic Ground Environment System.

Satcom Satellite communications

SCOOP, Project Scientific Computation of Optimum Problems

SCP Strategic Computing Program

SDC System Development Corporation

SDI Strategic Defense Initiative

SEAC Standards Eastern Automatic Computer

SEAL SEa, Air, Land. The U.S. Navy's commandos.

SEI Space Exploration Initiative. President Bush's plan to put humans on the moon by 2000 C.E. and on Mars by 2019 C.E.

SHF Super High Frequency

SINBAC System for Integrated Nuclear Battle Analysis Calculus

SRI Originally, Stanford Research Institute. Now that Stanford has severed itself from this think tank it is officially SRI.

SSN Ship, Submersible, Nuclear. Nuclear ballistic missile submarine.

Tacsats Tactical military satellites for intervening directly in battles

TACCS Tactical Army Combat Service Support Computer System

TRADOC Training document

UAV Unmanned Air Vehicle

UCSC University of California at Santa Cruz

UHF Ultra High Frequency

ULF Ultra Low Frequency

UNIVAC Universal Automatic Computer

USAF United States Air Force

USMC United States Marine Corp

USN United States Navy

USS United States Ship

VISTA Very Intelligent Surveillance and Target Acquisition Systems

VVAW Vietnam Veterans Against the War

WWMCCS Worldwide Military Command and Control System. Pronounced "Wimex."

WHIMPER Wound-Healing Injection Mandating Partial Early Recovery

References

The Past is Prologue.
   —*Motto on the National Archives Building*

In three sections: (1) Archives; (2) Books and Articles; (3) Government Documents.

1. Archives

National Archives of the United States

20th Air Force: Record Group 18, HQ 20th Air Force, Decimal File 1944-45. Box 8 (319.1); Box 9 (3.19.1 to 320.3); Box 16 (373.0); Box 17 (373.11); Box 18 (373.2 to 373.4)); Box 19 (380.01 to 385.0); Box 26 (No. Special to No. 3).

Records of the Office of the Chief of Naval Operations, Office of Naval Intelligence: Record Group. Box 8 (Foreign Intelligence Branch Tech. Section Op 23-F2, 1945-46). U-234 route slips.

Records of the Office of the Chief of Naval Operations: Record Group 28, Box 5 (OP-23 FN).

2. Books and Articles

(*Note:* All books published in New York City, New York, unless otherwise noted.)

Adams, Gordon. (1981). *The Iron Triangle: The Politics of Defence Contracting.* Council on Economic Priorities.

Aitken, H. (1985). *Scientific Management in Action: Taylorism at the Watertown Arsenal.* Princeton, N.J.: Princeton University Press.

Allen, Thomas. (1987). *War Games: The Secret World of the Creators, Players, and Policy Makers Rehearsing World War III Today.* McGraw-Hill.

Anderson, Edward. (1984). First Strike: Myth or Reality, in Paul Joseph and Simon Rosenblum, eds. *Search for Sanity: The Politics of Nuclear Weapons and Disarmament.* Boston: South End Press, pp. 130–137.

Andriole, Stephen, and Hopple, Gerald, eds. (1989). *Defense Applications of Artificial Intelligence*. Lexington, Mass.: Lexington Books.

Arar, Yardena. (1991). Now, Rob Lowe and Jan-Michael Vincent in *Desert Shield*, *San Jose Mercury*, March 15: E1.

Armstrong, David. (1982). *Bullets and Bureaucrats: The Machine Gun and the United States Army, 1861–1916*. Boulder, Colo.: Greenwood Press.

Arnold, Gen. H. H. (1946). Air Force in the Atomic Age, in Dexter Masters and Katherine Way, eds., *One World or None*. Whittsey House.

Arnold, Maj. Gen. Wallace C. (1995). Manprint: Battle Command and Digitalization, *Military Review*, May–June: 48–55.

Aronowitz, Stanley. (1988). Postmodernism and Politics, *Social Text* 50, Winter: 99–114.

——. (1988). *Science as Power: Discourse and Ideology in Modern Society*. Minneapolis: University of Minnesota Press.

Aronson, Robert. (1984). Robots Go To War, *Machine Design*, December 6: 73–78.

Arquilla, John, and Ronfeldt, David. (1993). Cyberwar is Coming! *Journal of Comparative Strategy*, 12, no. 2: 141–165.

Associated Press Service. (1987). Stark Captain Blames Deficient Radar for Frigate's Failure to Defend Itself, *San Jose Mercury News*, November 15: C8.

——. (1991a). No honors ceremonies for returning dead, *San Francisco Examiner*, January 20: A8.

——. (1991b). LA Policewoman Slain by Gunman, *Santa Cruz Sentinel*, February 12: A6.

Aviation Week & Space Technology Staff. (1985). Sandia, *Aviation Week & Space Technology*, March 11: 36.

——. (1986). February 17: 49.

Bakunin, Mikhail. (1953). *The Political Philosophy of Bakunin*, G. Maximoff, ed. Free Press.

Balzar, John. (1991). Video Horror of Apache Victims' Deaths, *Manchester Guardian*, February 25: 1.

Baranauskas, Tom. (1987). Fire Control on the Modern Battlefield, *Defense Electronics*, October: 137–144.

Baritz, Loren. (1985). *Backfire: Vietnam—American Culture and the Vietnam War*. Ballantine.

Barry, John, and Charles, Robert. (1992). Sea of Lies, *Newsweek*, July 12: 29–39.

Baudrillard, Jean. (1983). *Simulations*, Foss et al., trans. Semiotext(e).

——. (1991). The Reality Gulf, *The Guardian*, January 11: 25.

Baudrillard, Jean, and Lotringer, Sylvere. (1987). *Forget Foucault & Forget Baudrillard*, Beitchman et al., trans. Semiotext(e).

Bazerman, Charles. (1981). What Written Knowledge Does: Three Examples of Academic Discourse, *The Philosophy of the Social Sciences*, 11, no. 4, September: 361–389.

Beardon, Colin, ed. (1989). *Artificial Intelligence Terminology: A Reference Guide*. Wiley.

Beers, David. (1987). The Divided Mind of Artificial Intelligence, *Image Magazine*, February 15: 27–32, 38.

Bergen, John. (1985). *Military Communications—A Test For Technology: The U.S. Army in Vietnam*. Washington, D.C.: U.S. Army.

Bey, Hakim. (1995). The Information War, *CTheory* (www.ctheory.com), Electronic.

Biddle, Wayne. (1986). How Much Bang for the Buck?, *Discover*, September: 50–66.

Binkin, Martin. (1986). *Military Technology and Defense Manpower*. Washington, D.C.: Brookings Institution.

Blainey, Geoffrey. (1973). *The Causes of War*. Free Press.

Bond, Brian. (1987). Battlefield C³I Throughout History, *International Security* II: 125–129.

Bono, James. (1986). Literature, Literary Theory and the History of Science, *PSLS: Publication of the Society for Literature and Science* 2: 1–9.

———. (1991). Science, Discourse, and Literature: The Role/Rule of Metaphor in Science, in Stuart Peterfreund, ed. *Literature and Science: Theory and Practice*. Boston: Northeastern University Press.

Borland, C. (1967). *The Arts of the Alchemists*. Macmillan.

Bowdish, Lt. Cmdr. Randall G. (1995). The Revolution in Military Affairs: The Sixth Generation, *Military Review*, November–December: 26–33.

Bowman, Lt. Col. Robert. (1991). Iraq Bush-Whacked: Desert Storm Blows Over but Black Cloud Remains, *Space and Security News*, March 8: 1–11.

Brackney, Howard. (1959). The Dynamics of Military Combat, *Operations Research* 7, January–February: 30.

Brand, Stewart. (1974). *Two Cybernetic Frontiers*. Random House.

———. (1987). *The Media Lab: Inventing the Future at MIT*. Viking.

Broad, William. (1988). *Star Warriors*. Simon & Schuster.

Brodie, Bernard, ed. (1946). *The Absolute Weapon: Atomic Power and World Order*. Harcourt, Brace.

———. (1973). *War and Politics*. Macmillan.

Brodie, Bernard, and Brodie, Fawn. (1973). *From Crossbow to H-Bomb*. Bloomington: Indiana University Press.

Brown, Norman O. (1959). *Life Against Death: The Psychoanalytical Meaning of History*. Vintage.

———. (1966). *Love's Body*. Vintage.

Buck, Peter. (1985). Adjusting to Military Life: The Social Sciences Go to War, 1941–1950, in Merritt Row Smith, 1985a, pp. 203–252.

Bunker, Robert J. (1994). The Transition to Fourth Epoch War, *Marine Corp Gazette*, September: 23–32.

———. (1995a). Digital Battlefield Conference, *Military Review*, May–June: 3–4.

———. (1995b). Rethinking OOTW, *Military Review*, November–December: 34–41.

Burnham, David. (1983a). *The Rise of the Computer State*. Random House.

———. (1983b). More Power for Nation's Most Secretive Spy Agency, *San Francisco Chronicle*, April 6: A6.

Bush, Vannevar. (1949). *Modern Arms and Free Men*. Simon & Schuster.

Burke, Kenneth. (1945). *Grammar of Motives*. Prentice-Hall.

Business Week Staff. (1991a). The High-Tech War Machine, *Business Week*, February 4: 38–39.

———. (1991b). Managing the War, *Business Week*, February 4: 36–37.

————. (1991). This Army Marches on Silicon, *Business Week*, February 4: 40–43.

Byron, Cmdr. John. (1985). Warriors, *Proceedings—US Naval Institute*, June: 63–68.

Callaway, Enoch. (1976). Electrical "Windows" on the Mind: Applications for Neurophysiologically Defined Individual Differences, in Edward Salkovitz, ed., *Science, Technology, and the Modern Navy, 30th Anniversary 1946–1976*. Arlington, Va.: U.S. Navy, pp. 671–692.

Campbell, Joseph. (1962). *Creative Mythology*, Vol. 4: Masks of the Gods. Penguin Books.

Canan, James. (1975). *The Superwarriors: The Fantastic World of Pentagon Superweapons*. Weybright & Talley.

Caputi, Jane. (1987). *Age of Sex Crime*. Bowling Green, Ohio: Popular Press (Bowling Green State University).

Caputo, Philip. (1977). *A Rumor of War*. Ballantine.

Capuzzo, Mike. (1991). Once a Hawk, *San Jose Mercury News*, April 3: E12.

Cassott, Michael. (1988). Classified Astronauts, *Space World*, May: 6–8.

Catton, Bruce. (1981). *Reflections on the Civil War*. Berkley Books.

Center for Defense Information Staff, ed. (1983). *Quotes: Nuclear War*. Washington, D.C.: CDI.

Ceruzzi, Paul. (1983). *Reckoners: The Prehistory of the Digital Computer, from Relays to the Stored Program Concept, 1935–1945*. Westport, Conn.: Greenwood Press.

————. (1989a). *Beyond the Limits: Flight Enters the Computer Age*. Cambridge, Mass.: MIT Press.

————. (1989b). Electronics Technology and Computer Science 1940–1975: A Coevolution, *Annals of the History of Computing* 10, no. 4: 257–275.

Chaliand, Gerard. (1978). *Revolution in the Third World*, Johnson, trans. Penguin.

Chandor, Anthony, Graham, John, and Williamson, Robing, eds. (1981). *The Penguin Dictionary of Computers*. Penguin.

Chandler, David. (1974). *The Art of Warfare on Land*. London: Hamlyn Press.

Chapman, Gary. (1987a). Thinking About "Autonomous" Weapons, *CPSR Newsletter* 5, no. 3, Fall: 1, 11–14.

————. (1987b). The New Generation of High Technology Weapons, in David Bellin and Gary Chapman, eds., *Computers in Battle*. Harcourt Brace Jovanovich, pp. 61–100.

Childs, John Brown. (1991). Notes on the Gulf War, Racism, and African-American Social Thought: Ramifications for Teaching, *Journal of Urban and Cultural Studies* 2, no. 1: 81–92.

Chomsky, Noam. (1967). *American Power and the New Mandarins*. Vintage.

————. (1979). *Language and Responsibility*. Brighton, England: New Harvester.

————. (1986). The Rationality of Collective Suicide, *Canadian Journal of Philosophy* 12: 23–39.

————. (1989). The Fifth Freedom, *The Listener*, March 23: 6–9.

Chronicle of Higher Education Staff. (1994). Hot Type, March 23: A10.

Clancy, Tom. (1995). *Debt of Honour*. HarperCollins.

Clark, Don. (1991). How High Tech Gives US an Edge, *San Francisco Chronicle*, January 16: C1, C4.

Clarke, I. F. (1966). *Voices Prophesying War: 1763–1984*. London: Oxford University Press.

Clausewitz, Gen. Carl von (1962) *On War*, Graham, trans. Baltimore: Penguin.

Clodfelter, Mark. (1994). *The Limits of Air Power: The American Bombing of North Vietnam*. The Free Press.

Clynes, Manfred, and Kline, Nathan. (1960). Cyborgs and Space, *Astronautics*, September: 24–28.

Coates, James, and Kilian, Michael. (1984). *Heavy Loses: The Dangerous Decline of American Defense*. Penguin.

Cohen, I. Bernard. (1988). The Computer: A Case Study of Support By Government, Especially the Military, of a New Science and Technology, in M. Mendelssohn et al., eds., *Science, Technology and the Military—Sociology of the Sciences Yearbook 1988*, Vol. 1. Boston: Kluwer, pp. 119–154.

Cohn, Carol. (1987). Sex and Death in the Rational World of Defense Intellectuals, *Signs* 12, no. 4, Summer: 687–718.

Cooke, Miriam. (1991). Postmodern Wars: Phallomilitary Spectacles in the DTO, *The Journal of Urban and Cultural Studies* 2, no. 1: 27–40.

Cordes, Colleen. (1995). Pentagon Budget Gives Priority to Purely Military Research, *The Chronicle of Higher Education*, December 8: A26-7.

Corn, David. (1995). The Real Enemy, *The Nation* 250, no. 22: 781.

Coudert, Allison. (1980). *Alchemy—The Philosopher's Stone*. Boulder, Colo.: Shambala.

Counterspy Staff. (1974). Thought Control, *Counterspy* 2, Fall: 5.

Cronon, William. (1992). A Place for Stories: Nature, History, and Narrative, *The Journal of American History*, vol. 78, March: 1347–1375.

Crosby, Alfred. (1986). *Ecological Imperialism: The Biological Expansion of Europe, 900–1900*. Cambridge, England: Cambridge University Press.

Cross, Stephen, et al. (1986). Knowledge-Based Pilot Aids: A Case Study in Mission Planning, in W. Thomas and T. Wyner, eds., *Artificial Intelligence and Man–Machine Systems*. Springer-Verlag, pp. 141–174.

Crowther, J. and Whiddington, R. (1948). *Science at War*. Philosophical Library.

Crystal, Coca. (1972/1982) Airwar: Computerized Battlefield, in Dana Beal, ed., *The Secret History of the 70s*. Youth International Party, pp. 18–23.

Cunningham, Ann Marie, and Fitzpatrick, Mariana. (1983). *Future Fire: Weapons for the Apocalypse*. Warner Books.

Dallmayr, Fred. (1984). *Language and Politics*. South Bend, Ind.: University of Notre Dame Press.

Dasey, Charles. (1984). Biotechnology and Its Applications to Military Medical R&D, *Army Research, Development & Acquisition Magazine*, July–August: 12–13.

Davies, Owen. (1987). Robotic Warriors Clash in Cyberwars, *Omni*, January: 76–88.

Debray, Regis. (1991). Startling New Predictions: "The 21st Century Will Resemble the 19th," *The New Patriot* 4, no. 2 (March–April): 4.

Defense Electronics Staff. (1987). B-52Gs to Get Smart Weapons If Nuclear Treaty Passes, *Defense Electronics*, October: 11, 37.

———. (1987). *Technologies of Gender: Essays on Theory, Film, and Fiction*. Bloomington: Indiana University Press.

Deitchman, Seymour. (1979). *New Technologies and Military Power*. Boulder, Colo.: Westview Press.

———. (1985). Weapons, Platforms, and the New Armed Services, *Issues in Science and Technology*, Spring: 83–107.

De Landa, Manuel. (1991). *War in the Age of Intelligent Machines*. Zone Books.

De Lauretis, Teresa. (1984). *Alice Doesn't: Feminism, Semiotics, Cinema*. Bloomington: Indiana University Press.

Deleuze, Gilles, and Guattari, Félix. (1986). *Nomadology: The War Machine*, B. Massumi, trans. Semiotext(e).

Dennet, Daniel. (1981). *Brainstorms: Philosophical Essays on Mind and Psychology*. Cambridge, Mass.: MIT Press.

Der Derian, James. (1990). The (S)pace of International Relations: Simulation, Surveillance, and Speed, *International Studies Quarterly* 34: 295–310.

————. (1991). Cyberwar, Video Games, and the New World Order, Second Annual Cyberspace Conference, Santa Cruz, Calif., April.

Derrida, Jacques. (1984). Full Speed Ahead, Seven Missiles, Seven Missives, *diacritics* 14, no. 2, Summer: 20–31.

De Seversky, Maj. Alexander P. (1942). *Victory Through Air Power*. Simon & Schuster.

Deutsche Forschung-und Versuchsanstalt, eds. (1986). *Lecture Notes in Control and Information Sciences*. Springer-Verlag.

Dickson, David. (1984). *The New Politics of Science*. Pantheon.

Dickson, Paul. (1971). *Think Tanks*. Atheneum.

————. (1976). *The Electronic Battlefield*. Atheneum.

Dillon, G. M. (1988). Defense, Discourse and Policy Making: Britain and the Falklands, Working Paper no. 4, First Annual Conference on Discourse, Peace, Security and International Society, Institute for Global Conflict and Cooperation, University of California at San Diego.

Din, Allen, ed. (1987). *Arms and Artificial Intelligence: Weapon and Arms Control Applications of Advanced Computing*. London: Sipri/Oxford Press.

Dinnerstein, Dorothy. (1977). *The Mermaid and the Minotaur*. Harper & Row.

Dixon, Norman. (1976). *On the Psychology of Military Incompetence*. Basic Books.

Dobos, Manuela. (1993). Rape: War Crime Against Women and a Nation, *Peace and Democracy News*, vol. 7, no. 1, Summer: 16–19.

Donner, Frank. (1980). *The Age of Surveillance: The Aims and Methods of America's Political Intelligence System*. Vintage Books.

Donnini, Lt. Col. Frank. (1990). Douhet, Caproni and Early Air Power, *Air Power History* 37, no. 2, Summer: 45–51.

Donovan, Col. James. (1970). *Militarism U.S.A.* Scribner's.

Dover, R. (1991). New Military Doctrines, *Israel & Palestine*, March: 13.

Dower, John. (1986). *War Without Mercy: Race & Power in the Pacific War*. Pantheon.

Dreyfus, Hubert. (1979). *What Computers Can't Do*. Harper Colophon.

Dreyfus, Hubert, and Dreyfus, Stuart. (1985). *Mind Over Machine*. Free Press.

Dreyfus, Hubert, and Rabinow, Paul. (1983). *Michel Foucault: Beyond Structuralism and Hermeneutics*. Chicago: University of Chicago Press.

Druzhinin, V. V. and Kontorov, D. S. (1975). *Concept Algorithm, Decision*, Washington, D.C.: U.S. Government Printing Office.

Dyer, Gwynne. (1985). *WAR*. Crown.

Dyson, Freeman. (1984). *Weapons and Hope*. Harper & Row.

Easlea, Brian. (1980). *Witch Hunting, Magic and the New Philosophy: An Introduction to Debates of the Scientific Revolution 1450–1750.* Brighton, England: New Harvester.

Easterbrook, Gregg. (1991). "High Tech" Isn't Everything, *Newsweek,* February 18: 49.

Editors of Time–Life. (1988). *Understanding Computers: The Military Frontier.* Alexandria, Va.: Time–Life.

Edwards, Paul. (1986a). *Technologies of the Mind—Computers, Power, Psychology, and World War II.* Working Paper no. 2, Silicon Valley Research Group, University of California at Santa Cruz.

———. (1986b). Border Wars: The Science and Politics of Artificial Intelligence, *Radical America* 19, no. 6: 39–50.

———. (1986c). Artificial Intelligence and High Technology War: The Perspective of the Formal Machine, Working Paper no. 6, Silicon Valley Research Group, University of California at Santa Cruz.

———. (1987). A History of Computers and Weapons Systems, in David Bellin and Gary Chapman, eds., *Computers in Battle.* Harcourt Brace Jovanovich, pp. 45–61.

———. (1989). The Closed World: Systems Discourse, Military Policy and Post-World War II US Historical Consciousness, in L. Levidow and K. Robins, eds., *Cyborg Worlds: The Military Information Society.* London: Free Association Press, pp. 135–158.

———. (1991). The Army and the Microworld: Computers and the Militarized Politics of Gender identity, *Signs* 16, no. 1, Fall: 102–27.

———. (1996). *The Closed World.* Cambridge, Mass: MIT Press.

Eglash, Ron. (1989). A Cybernetics of Chaos, Qualifying Essay for the History of Consciousness Board of Studies, Oakes College, University of California at Santa Cruz.

Ehrenreich, Barbara. (1987a). Iranscam: The Real Meaning of Oliver North—Can a Member of the Warrior Caste Survive the Horror of Peace? *Ms.* May: 26.

———. (1987b). Foreword, in Klaus Theweleit, *Male Fantasies,* Vol. 1: Women Floods, Bodies History. Minneapolis: University of Minnesota Press, pp. ix–xxii.

———. (1991). War for War's Sake, *Z Magazine* March: 23–25.

Ekstein, Modris. (1989). *Rites of Spring: The Great War and the Birth of the Modern Age.* Boston: Houghton Mifflin.

Eldridge, Bo. (1991). Desert Storm: Mother of All Battles, *Command* 13, November–December: 12–42.

Eliade, Mircea. (1962). *The Forge and the Crucible.* Harper & Row.

Ellis, John. (1973). *The Social History of the Machine Gun.* Pantheon Books.

Ellsberg, Daniel. (1972). *Papers on the War.* Simon & Schuster.

Emerson, Gloria. (1972, 1985) *Winners and Losers: Battles, Retreats, Gains, Losses, and Ruins from the Vietnam War.* Penguin.

Enloe, Cynthia. (1983). *Does Khaki Become You? The Militarisation of Women's Lives.* London: Pluto Press.

———. (1993). *The Morning After: Sexual Politics and the End of the Cold War.* Berkeley: University of California Press.

Epstein, Barbara. (1986). The Culture of Direct Action: The Livermore Action Group and the Peace Movement, *Socialist Review* 82/83: 43–52.

Falk, Jim. (1987). The Discursive Shaping of Technological Change: The Case of Nuclear War, paper presented at the 1987 Meeting of the Society for the Social Studies of Science, Worcester, Mass.

Fallows, James. (1982). *National Defense*. Vintage.

Faludi, Susan. (1986). The Billion Dollar Toy Box, *West Magazine, Sunday San Jose Mercury*, November 21: 12–18.

Farhat, Laina. (1990). Computer Analysis Tells a Grim Tale of War, *San Francisco Chronicle*, October 23: A21.

Fee, Elizabeth. (1986). Critiques of Modern Science: The Relationship of Feminism to Other Radical Epistemologies, in R. Bleier, ed., *Feminist Approaches to Science*. Pergamon Press.

Feigenbaum, E., and Feldman, J., eds. (1963). *Computers and Thought*. McGraw Hill.

Ferguson, Kathy. (1977). *A Feminist Critique of Bureaucracy*. Philadelphia: Temple University Press.

Ferrill, Arthur. (1985). *The Origins of War: From the Stone Age to Alexander the Great*. London: Thames & Hudson.

FitzGerald, Frances. (1972). *Fire in the Lake: The Vietnamese and the Americans in Vietnam*. Boston: Little Brown.

Fjermedal, Brant. (1990). *The Tomorrow Makers: A Brave New World of Living-Brain Machines*. Macmillian.

Flamm, Kenneth. (1987). *Targeting the Computer: Government Support and International Competition*. Washington, D.C.: Brookings Institution.

Flax, Jane. (1987). Postmodernism and Gender Relations in Feminist Theory, *Signs* 12, no. 4, Summer: 621–643.

Floyd, Christine. (1985). The Responsible Use of Computers: Where Do We Draw the Line? *CPSR Newsletter*, June: 1, 4–7.

Foltz, Kim. (1990). Army Uses Computer Disks to Recruit Video Generation, *New York Times*, March 21: C19.

Ford, Daniel. (1985). *The Button*. Simon & Schuster.

Forester, Tom, ed. (1981). *The Microelectronics Revolution*. Cambridge, Mass.: MIT Press.

———. (1985). *The Information Technology Revolution*. Oxford, England: Basil Blackwell.

Foucault, Michel. (1972). *The Archeology of Knowledge*. J. Smith, trans. Pantheon.

———, ed. (1975). *I, Pierre Riviere Having Slaughtered My Mother, My Sister, and My Brother . . .* , E. Jellinek, trans. Lincoln: University of Nebraska Press.

———. (1977). *Discipline and Punish: The Birth of the Prison*, P. Sheriden, trans. Vintage.

———. (1980). *Power/Knowledge: Selected Interviews and Other Writings, 1972–1977*, Gordon et al., trans. Pantheon Books.

———. (1984). *The Foucault Reader*, P. Rabinow, ed., Pantheon.

———. (1986). Nietzche, Genealogy, History, in Paul Rabinow, ed., *The Foucault Reader*. London: Houndsworth/Penguin.

Franklin, Bruce. (1988). *War Stars: The Superweapon and the American Imagination*. Oxford: Oxford University Press.

Frantz, Douglas. (1991). Big Role in Gulf for Pilotless Aircraft, *San Francisco Chronicle*, February 11: A9.

Fraser, Cmdr. Powell. (1988). The Out and Down Revolutions, *Proceedings—US Naval Institute*, December: 53–57.

Frederick, Donald. (1991). How Spies in the Sky Saw Desert Storm, *San Francisco Chronicle*, March 27: 2–3.

Friedman, Norman. (1989). The *Vincennes* Incident, *Proceedings—US Naval Institute*, May: 72–7.

———. (1991). *Desert Victory: The War for Kuwait*. Annapolis Md.: Naval Institute Press.

Friedman, Col. Richard, et al. (1982). *Advanced Technology Warfare*. Harmony Books.

Friedrich, Otto. (1991). *Desert Storm: The War in the Persian Gulf*. Richmond, Va.: Time.

Frye, Northrop. (1957). *Anatomy of Criticism*. Princeton, N.J.: Princeton University Press.

Fulghum, David. (1991). Allies Divide Air Strike Targets Into Grid of "Killing Boxes," *Aviation Week & Space Technology*, February 18: 62.

Fuller, Maj. Gen. J. F. C. (1943). Machine Warfare, *The Infantry Journal*: 60–67.

Fussell, Paul. (1975). *The Great War and Modern Memory*. Oxford: Oxford University Press.

———. (1989a). *Wartime: Understanding and Behavior in the Second World War*. Oxford: Oxford University Press.

———. (1989b). Conversations, *The United States Institute of Peace Journal*, September: 1, 8.

Gabriel, Richard. (1985). *Military Incompetence: Why the American Military Doesn't Win*. Toronto: Collins.

———. (1987). *No More Heroes: Madness & Psychiatry in War*. Hill & Wang.

Galbraith, John Kenneth. (1969a). How to Control the Military? *Harper's*, June: 32–33.

———. (1969b). Scaring the Hell Out of Everybody, *The Progressive*, June: 15.

———. (1981). *A Life in Our Times: Memoirs*. Boston: Houghton Mifflin.

Garchik, Leah. (1991). War Was Healthy and So Are Sardines, *San Francisco Chronicle*, March 18: A8.

Gellman, Barton. (1994). Revisiting the Gulf War, *Washington Post*, September 7: A1, A20.

Geertz, Clifford. (1973). *Interpretation of Culture*. Basic Books.

Generals (Brig. M. Harbottle et al.). (1984). *Generals for Peace and Disarmament*. Universe Books.

Gerth, Jeff. (1989). Updating of Pentagon Computers Runs Over Budget, Report Finds, *New York Times*, December 1: A20.

Gessen, Masha. (1995). Balkans Online, *Wired*, November: 156–162, 220–223.

Giap, Gen. Vo Nguyen. (1970). *The Military Art of People's War*. Monthly Review Press.

Gibson, James. (1986). *The Perfect War: Technowar in Vietnam*. Boston: Atlantic Monthly Press.

Giedion, Siegfried. (1948). *Mechanization Takes Command*. Oxford: Oxford University Press.

Gilbert, Felix. (1943). Machiavelli: The Renaissance of the Art of War, in Edward

Earle, ed., *Makers of Modern Strategy: Military Thought from Machiavelli to Hitler*. Princeton, N.J.: Princeton University Press.

Gill, Karamjit, ed. (1986). *Artificial Intelligence for Society*. Wiley.

Gilmartin, Trish. (1987). SDI Battle Management Slighted, Experts Charge, *Defense News*, September 28: 11–26.

Glass, Frank. (1991). US May Win the War, Lose Peace, *Santa Cruz Sentinel*, January 24: A8.

Glover, Michael. (1982). *The Velvet Glove: The Decline and Fall of Moderation in War*. London: Hodder & Stoughton.

Goben, Ron. (1987). Human Nerve Repair May Be Possible, *Stanford Observer*, Spring: 12–13.

Golden, Tim. (1994). The Voice of the Rebels Has Mexicans in His Spell, *New York Times*, February 8: A3.

Gomez, Dr. Richard, and Van Atta, Capt. Michael. (1984). AirLand Battlefield Environment Thrust, *Army RD&A Magazine*, May–June: 17.

Goodman, Glenn W. Jr. (1989). Vincennes Tragedy Could Be Repeated Closer to Home, *Armed Forces Journal International*, July: 79–80, 83.

Goodman, Paul. (1964). *Utopian Essays and Practical Proposals*. Vintage.

Gordon, Don. (1987). The JTIDS/PLRS Hybrid—A NATO Standard? *Military Technology* 11, no. 5: 97–106.

Gordon, Michael R., and Trainor, Gen. Bernard E. (1995). *The General's War: The Inside Story of the Conflict in the Gulf*. Little, Brown.

Gorn, Michael. (1988). *Harnessing the Genie: Science and Technology Forecasting for the Air Force, 1944–1986*. Washington, D.C.: U.S. Air Force.

Gould, Stephen J. (1987). *Time's Arrow/Time's Cycle: Myth and Metaphor in the Discovery of Geological Time*. Cambridge, Mass.: Harvard University Press.

Graham, Gen. Daniel. (1983). *High Frontier: There Is a Defense Against Nuclear War*. Tor Books.

Gray, Chris Hables. (1989). The Cyborg Soldier: The U.S. Military and the Post-modern Warrior, in L. Levidow and K. Robins, eds., *Cyborg Worlds: Programming the Military Information Society*, London: Free Association Press; New York: Columbia University Press, pp. 43–73.

———. (1993). The Culture of War Cyborgs: Technoscience, Gender, and Post-modern War, *Research in Philosophy & Technology*, special issue on technology and feminism, vol. 13, Joan Rothschild, ed.: 141–163.

———. (1994). "There Will Be War!" Future War Fantasies and Militaristic Science Fiction in the 1980s, *Science-Fiction Studies* 21, no. 64, pt. 3, November: 302–314.

———. (1997). AI at War: An Analysis of the Aegis System in Combat, in D. Schuler, ed., *Directions and Implications of Advanced Computing 1990*, Vol. 3, Ablex, pp. 62–79.

Gray, J. Glenn. (1959). *The Warriors: Reflections on Men in Battle*. Harper & Row.

Green, Charles, and Greve, Frank. (1991). In This War, Civilians Faring Better, *San Jose Mercury News*, February 15: A19.

Greve, Frank. (1991). US Troops Low on Ammunition, *San Jose Mercury News*, February 13: A1, A15.

Grier, Peter. (1986a). Congress Goes after Suspicious Weapons-Test Results, *Christian Science Monitor*, 19 May: 3–4.

———. (1986b). Aegis to Put Swagger in the Navy's Step, *Christian Science Monitor*, August 21: 3.

Griffin, David, ed. (1988). *The Reenchantment of Science*. Albany: State University of New York Press.

Griffin, Susan. (1978). *Woman and Nature: The Roaring Inside Her*. Harper & Row.

Griffith, Col. Samuel B. (1962). Introduction in Sun Tzu, *The Art of War*, S. B. Griffith, trans. Oxford: Oxford University Press, pp. 1–56.

Guice, Jon. (1994). *Designing the Future: Next-Generation Computing and the US Advanced Research Projects Agency*, Dissertation, Sociology Department, University of California at San Diego.

Guillermoprieto, Anna. (1995). Letter from Mexico: The Unmasking, *The New Yorker*, March 13: 40–48.

Gusfield, Joseph. (1976). The Literary Rhetoric of Science: Comedy and Pathos in Drinking Driver Research, *American Sociological Review* 41, February: 16–34.

Gust, Kelly. (1990). Computers Smashed to Prevent World War, Says Anti-War Activist, *Peninsula Times Tribune*, December 9: A9.

Gusterson, Hugh. (1987). *Rationality and Transgression in the Search for a Post-modern Politics of Resistance*, paper presented at the American Anthropology Association Meeting, Chicago, Ill.

———. (1991). Nuclear War, the Gulf War, and the Disappearing Body, *Journal of Urban and Cultural Studies* 2, no. 1: 45–56.

Hacking, Ian. (1983). *Representing and Intervening*. Cambridge, England: Cambridge University Press.

Halberstam, David. (1972). *The Best and the Brightest*. Fawcett Crest.

Hales, Mike. (1974). Management Science and "The Second Industrial Revolution," *Radical Science Journal* 1, January: 5–28.

Hall, Alfred Rupert. (1952/1969) *Ballistics in the 17th Century*. Cambridge, England: Cambridge University Press.

Hanson, Victor. (1989). *The Western Way of War: Infantry Battle in Classical Greece*. Knopf.

Haraway, Donna. (1981–1982) The High Cost of Information in Post-World War II Evolutionary Biology, *Philosophical Forum* 13, no. 2/3: 244–278.

———. (1985). A Manifesto for Cyborgs: Science, Technology, and Socialist Feminism for the 1980s, *Socialist Review* 80: 95–107.

———. (1988). Situated Knowledges: The Science Question in Feminism and the Privilege of Partial Perspective, *Feminist Studies* 14, no. 3, Fall: 575–599.

Harding, Sandra. (1986). *The Science Question in Feminism*. Ithaca, N.Y.: Cornell University Press.

Harris, Robert, and Paxman, Jeremy. (1982). *A Higher Form of Killing: The Secret Story of Chemical and Biological Warfare*. Hill & Wang.

Haugeland, John. (1987). *Artificial Intelligence: The Very Idea*. Cambridge, Mass.: MIT Press.

Hawley, T.M. (1992). *Against the Fires of Hell: The Environmental Disaster of the Gulf War*. Harcourt Brace Jovanovich.

Headrick, Daniel. (1981). *The Tools of Empire: Technology and European Imperialism in the Nineteenth Century*. Oxford University Press.

Healy, Melissa. (1991). Stealthy Commandos Provided Vital Information, *San Jose Mercury News*, Februrary 28: 12A.

Heims, Steven. (1980). *John von Neumann and Norbert Wiener: From Mathematics to the Technologies of Life and Death*. Cambridge, Mass.: MIT Press.

Heiser, Jon, et al. (1979). Can Psychiatrists Distinguish a Computer Program Simulation of Paranoia from the Real Thing? The Limitations of Turing-like Tests as Measures of the Adequacy of Simulations, *Journal of Psychiatric Research* 15, no. 3: 149–62.

Hellman, Peter. (1987). The Little Airplane That Could, *Discover*, February: 78–87.

Henderson, Col. Darryl. (1991). How Army Is Marketing the Gulf War to "Soft" Public, *San Francisco Examiner*, February 10: A8.

Herken, Gregg. (1985). *Counsels of War*. Knopf.

Herr, Michael. (1978). *Dispatches*. Avon.

Hersh, Seymour. (1987). How the US Plotted to Kill Gadhafi, *San Jose Mercury News*, February 22: 1.

Hilgartner, Stephen, et al. (1982). *Nukespeak: The Selling of Nuclear Technology in America*. San Francisco: Sierra Club Books.

Hilts, Philip, and Moore, Molly. (1988). The "Shield of the Fleet" May Have a Few Holes in It, *Washington Post National Weekly Edition*, July 11–17: 8.

Hodges, Andrew. (1983). *Alan Turing: The Enigma*. Touchstone Books.

Hofstadter, Albert. (1955). The Scientific and Literary Uses of Language, in D. Bryson and R. Lyman et al., eds., *Symbols and Society*. Simon and Schuster.

Holley, Irving B., Jr. (1983). *Ideas and Weapons: Exploitation of the Aerial Weapon by the United States During World War I*. Washington, D.C.: Office of Air Force History. (Originally published 1953, New Haven, Conn.: Yale University Press)

———. (1969). The Evolution of Operations Research and Its Impact on the Military Establishment: The Air Force Experience, in Lt. Col. Monte Wright and Lawrence Paszek, eds., *Science, Technology, and Warfare: The Proceedings of the Third Military History Symposium*. Washington, D.C.: U.S. Government Printing Office, pp. 110–121.

———. (1988). Doctrine and Technology as Viewed by Some Seminal Theorists of the Art of Warfare from Clausewitz to the Mid-Twentieth Century, in Stuart Pfaltzgraff, Jr., ed., *Emerging Doctrines and Technologies: Implications for Global and Regional Political–Military Balances*. Lexington, Mass.: Lexington Books, pp. 12–34.

Holmes, Richard. (1986). *Acts of War—The Behavior of Men in Battle*. Free Press.

Homer-Dixon, Thomas. (1987). A Common Misapplication of the Lanchester Square Law, *International Security* 12, no. 1, Summer: 135–138.

Howard, Michael. (1984). *The Causes of Wars and Other Essays*. Cambridge, Mass.: Harvard University Press.

Hoy, David. (1982). *The Critical Circle*. Berkeley: University of California Press.

———, ed. (1986). *Foucault: A Critical Reader*. Oxford: Basil Blackwell.

Hudgens, Gerald, and Holloway, W. (1971). *Behavior Modification and Changes in Central Nervous System Biochemistry: An Annotated Bibliography*. Aberdeen Proving Ground, Md.: Human Engineering Laboratories.

Hunt, James, and Blair, John, eds. (1985). *Leadership on the Future Battlefield*. McClean, Va.: Pergamon-Brassey's.

Hupp, Brian. (1987). Strategic Dissemination of Information—Livermore Tests SDI, *City on a Hill*, May 15: 7.

Hynes, Samuel. (1991). The Death of Landscape, *MHQ: The Quarterly Journal of Military History*, Spring: 17–27.

IEEE Spectrum Staff. (1984). *IEEE Spectrum* 19, no. 8.

Inglis, Paul. (1992). *The Cruel Peace: Everyday Life and the Cold War*. Basic Books.

Isokoff, Michael, and Brown, Warren. (1987). Nothing Like a Treaty to Boost the Arms Business, *Washington Post National Weekly Edition*, December 21: 21.

Isokoff, Michael. (1990a). War on Drugs Mobilizes National Guard, *Washington Post*, August 14: A1, A4.

———. (1990b). Interest in Grateful Dead Was Not Musical, *Washington Post*, August 14: A4.

James, William. (1911). *Memories and Studies*. Longmans, Green.

Jameson, Fredric. (1984). Postmodernism, or on the Cultural Logic of Late Capitalism, *New Left Review* 146, July–August: 53–92.

Janis, Irving. (1972). *Victims of Groupthink: A Psychological Study of Foreign Policy Decisions and Fiascoes*. Boston: Houghton Mifflin.

Janowitz, Morris. (1971). *The Professional Soldier: A Social and Political Portrait*. Free Press.

Jervis, Robert, et al. (1985). *Psychology and Deterrence*. Baltimore: Johns Hopkins University Press.

Johns, Lt. Cmdr. Eric. (1988). Perfect is the Enemy of Good Enough, *Proceedings—US Naval Institute*, October: 37–48.

Johnson, David. (1988). Panel Reports Ethics Conflicts in SDI Project, *New York Times*, April 23: 1.

Johnson, Steve. (1986). Futuristic Defense is Studied, *San Jose Mercury News*, December 5: A1, A17.

———. (1987). Superconductors May Be Key to Superweapons, *San Jose Mercury News*, August 26: A1, A24.

———. (1988). Flaws Detailed in Blue Cube's Computer, *San Jose Mercury News*, August 13: B1, B4.

Kaldor, Mary. (1981). *The Baroque Arsenal*. Hill & Wang.

———. (1986). The Weapons Succession Process, *World Politics* 37, no. 4, July: 577–595.

———. (1987). The Imaginary War, in B. Smith and E. P. Thompson, eds., *Prospectus for a Habitable Planet*. Penguin.

Kane, Col. Francis. (1964). Security Is Too Important to Be Left to Computers, *Fortune*, April: 146–147, 231–236.

Kaplan, Fred. (1983). *Wizards of Armageddon*. Touchstone Books.

Karch, Col. Lawrence. (1988). The Corps in 2001, *Proceedings—US Naval Institute*, November: 37–41.

Karras, Thomas. (1983). *The New High Ground: Strategies and Tactics of Space Age War*. Simon & Schuster.

Karush, Fred. (1986). Metaphor in Immunology, *Science News*, October 19: 254.

Keegan, John. (1976). *The Face of Battle*. Penguin.

————. (1987). *The Mask of Command*. Viking.

————. (1993). *A History of Warfare*. Knopf.

Keen, Cmdr. Timothy. (1988). Artificial Intelligence and the 1,200-Ship Navy, *Proceedings—US Naval Institute*, October: 96–100.

Keenan, Faith. (1991). No Tally of Enemy Dead: US Refuses to List Iraq Casualties, *San Francisco Examiner*, January 20: A10.

Keithly, Cmdr. Thomas. (1988). Tomorrow's Surface Forces, *Proceedings—US Naval Institute*, December: 50–61.

Keller, Evelyn Fox. (1985). *Reflections on Gender and Science*. New Haven, Conn.: Yale University Press.

Kellner, Douglas. (1997). From Nam to the Gulf: Postmodern Wars?, in Michael Bibby, ed., *The Viet Nam War and Postmodernity*. Next Generation.

Kennan, George. (1961). *Russia and the West Under Lenin and Stalin*. Boston: Little, Brown.

Kennedy, Paul. (1987). *The Rise and Fall of the Great Powers*. Random House.

Kennedy, William. (1987). Why America's National-Security Planning Process Went Awry: Military Mentality Took Control When Checks and Balances Failed, *Christian Science Monitor*, January 12: 8.

Kennett, Lee. (1982). *A History of Strategic Bombing*. Scribner's.

Kevles, Daniel. (1987). *The Physicists: The History of a Scientific Community in Modern America*. Cambridge, Mass.: Harvard University Press.

Klare, Michael. (1972). *War Without End: American Planning for the Next Vietnams*. Knopf.

————. (1990). Policing the Gulf—and the World, *The Nation*, October 15: 416.

Klare, Michael. (1995). The New "Rogue State" Doctrine, *The Nation* 260, no. 18, May 8: 625–628.

Klare, Michael, and Kornbluth, Peter, eds. (1987). *Low Intensity Warfare: Counter-insurgency, Proinsurgency, and Antiterrorism in the Eighties*. Pantheon.

Klein, Jeffrey. (1991). Gulf War: Mother of "Star Wars," *San Jose Mercury News*, March 10: C1, C8.

Klossowski, Stanislas. (1973). *The Secret Art of Alchemy*. Avon Books.

Knightley, Philip. (1975). *The First Casualty: The War Correspondent as Hero, Propagandist, and Myth Maker*. Harvest Press.

Kochetkov, Gennady, Averchev, Vladimir, and Sergeev, Viktor. (1987). Artificial Intelligence and Disarmament, in Din, 1987, pp. 153–160.

Kranzberg, Melvin. (1969). Science, Technology and Warfare: Action, Reaction, and Interaction in the Post-World War II Era, in Lt. Col. Monte Wright and Lawrence Paszek, eds., *Science, Technology, and Warfare: The Proceedings of the Third Military History Symposium*. Washington, D.C.: U.S. Government Printing Office.

Krepinevich, A., Jr. (1986). *The Army and Vietnam*. Baltimore: Johns Hopkins University Press.

Kropotkin, Peter. (1970). *Kropotkin's Revolutionary Pamphlets*, R. Baldwin, ed. Dover.

Kuhn, Thomas. (1970). *The Structure of Scientific Revolutions*. Chicago: University of Chicago Press.

Kull, Steven. (1988). *Minds at War*. Basic Books.

Ladd, John. (1987). Computers at War: Philosophical Reflections on Ends and

Means, in D. Bellin and G. Chapman, eds., *Computers in Battle*, Harcourt Brace Jovanovich, pp. 286–312.

Lakoff, George, and Johnson, Mark. (1980). *Metaphors We Live By*. Chicago: University of Chicago Press.

Lakatos, I., and Musgrave, A., eds. (1970). *Criticism and the Growth of Knowledge*. Cambridge, England: Cambridge University Press.

Lamb, David. (1991). Reflections of "Norman of Arabia," *San Jose Mercury News*, February 26: 14A.

Landau, Susan. (1987). The Responsible Use of "Expert" Systems, paper presented at DIAC-87 and in *Symposium Proceedings*. CPSR, Inc., pp. 167–182.

Lantham, Donald. (1987). C³I Acquisition Strategies, *Military Technology* 5: 18–21.

Largelle, Orin. (1995). One of America's Last Tropical Rain Forests Under Siege, *Earth First Journal*, Spring: 2–4.

Larsen, Knud. (1986). Social Psychological Factors in Military Technology and Strategy, *Journal of Peace Research* 23, no. 4: 391–398.

Latour, Bruno. (1981). Insiders and Outsiders in the Sociology of Science, *Past and Present* 3: 199–216.

―――. (1987). *Science in Action*. Cambridge, Mass.: Harvard University Press.

―――. (1991). The Impact of Science Studies on Political Philosophy, *Science, Technology & Human Values* 16, no. 1, Winter: 3–19.

Latour, Bruno, and Woolgar, Steve. (1979). *Laboratory Life: The Social Construction of Scientific Facts*. Beverly Hills, Cal.: Sage Publications.

Lawrence, T. E. (1936). *Seven Pillars of Wisdom: A Triumph*. Doubleday.

Lebow, Richard Ned, and Stein, Janice Gross. (1994). *We All Lost the Cold War*. Princeton, N.J.: Princeton University Press.

Lederer, Edith M. (1995). From Angola to Yugoslavia, Women Are Waging Peace, *Oregonian*, August 28, pp. A1, A6.

Lepinwell, John. (1987) The Laws of Combat?, *International Security* 12, no. 1, Summer: 89–134.

Levidow, Les, and Robins, Kevin, eds. (1989). *Cyborg Worlds: The Military Information Society*. London: Free Association Press.

Levy, Steven. (1984). *Hackers: Heroes of the Computer Revolution*. Dell.

Lifton, Robert Jay. (1970). *Boundaries: Psychological Man in Revolution*. Touchstone Books.

―――. (1987). *The Future of Immorality and Other Essays for a Nuclear Age*. Basic Books.

Lifton, Robert Jay, and Falk, Richard. (1982). *Indefensible Weapons: The Political and Psychological Case Against Nuclearism*. Basic Books.

Lifton, Robert Jay, and Humphrey, Nicholas, eds. (1984). *In a Dark Time—Images for Survival*. Cambridge, Mass.: Harvard University Press.

Lilienfeld, E. (1979). *The Rise of Systems Theory*. Wiley.

Lind, William. (1985). A Doubtful Revolution, *Issues in Science and Technology*, Spring: 109–112.

Lindberg, James. (1984). The Army's New Thrust Initiative—A New Way of Doing Business, *Army RD&A Magazine*, Sept.–Oct.: 23–25.

Livingstone, Neil. (1990). *The Cult of Counterterrorism*. Lexington, Mass.: Lexington Books.

Loeb, Paul. (1986). *Nuclear Culture: Living and Working in the World's Largest Atomic Complex*. Santa Cruz, Cal.: New Society.

Loevinger, Lee. (1989). Book Reviews, *Jurimetrics: Journal of Law, Science and Technology* 29, no. 3, Spring: 359–364.

Longino, Helen. (1990). *Science as Social Knowledge: Values and Objectivity in Scientific Inquiry*. Princeton, N.J.: Princeton University Press.

Lovece, Joseph. (1991). Big Plans, High Hurdles for UAV Joint Program Office, *Armed Forces Journal International*, July: 45–49.

Luttwak, Edward. (1971). *A Dictionary of Modern War*. Harper & Row.

Lutz, Catherine. (1995). *The Psychological Ethic and the Spirit of Permanent War: The Production of Twentieth-Century American Subjects*, unpublished manuscript, University of North Carolina at Chapel Hill.

Lyotard, Jean-François. (1984) *The Postmodern Condition: A Report on Knowledge*. Geoff Bennington and Brian Massumi, trans. Minneapolis: University of Minnesota Press.

———. (1987). Rules and Paradoxes and Svelte Appendix, *Cultural Critique* 4: 209–219.

Machiavelli, Niccolò (1977) *The Prince*, Robert M. Adams, trans. Norton.

———. (1990). *The Art of War*, E. Farnsworth, trans. Da Capo Press.

Mackenzie, Angus. (1987). Reagan Implants a Network of Secrecy in Government, *San Jose Mercury News*, August 4: 7B.

Macksey, Kenneth. (1986). *Technology in War: The Impact of Science on Weapon Development and Modern Battle*. Prentice Hall.

Makiya, Kanan. (1993). Rape in the Service of the State, *The Nation*, May 10: 624–627.

Manno, Jack. (1984). *Arming the Heavens: The Hidden Military Agenda for Space, 1945–1995*. Dodd, Mead.

Mansfield, Susan. (1982). *The Gestalt of War: An Inquiry into Its Origin and Meaning as a Social Institution*. Dial Press.

Manzione, Elton. (1986). The Search for the Bionic Commando, *The National Reporter* 10, Fall/Winter: 36–38.

Marcos, Subcommandante (1994). Interview. *Love and Rage*, August: 9-14.

Marcus, Daniel. (1987). Pentagon Crisis Center Under Construction, *Defense News*, September 28: 9.

Marcus, Daniel, and Leopold, George. (1987). Growing Pains Accompany Modern Warfare's C[3] Program Expansion, *Defense News*, September 28: 11–26.

Markoff, John. (1986). Pentagon Battles Computer Information Explosion, *San Francisco Examiner*, December 28: 1.

Marks, John. (1979). *The Search for the "Manchurian Candidate": The CIA and Mind Control*. New York Times Books.

Markusen, Ann, and Yudkin, Joel. (1992). *Dismantling the Cold War Economy*. Basic Books.

Marshall, S. L. A. (1947). *Men Against Fire*. Morrow.

Martin, Laurence. (1983). *The Changing Face of Nuclear Warfare*. Harper & Row.

Martinez, Don. (1990). Peace Activist Guilty of Vandalism, *San Francisco Examiner*, December 14: A13.

Masterson, Rear Adm. Kleber, Jr. and Tritten, Cmdr. James. (1987). New Concepts in Global War-gaming, *Proceedings—US Naval Institute*, July: 117–119.

McGucken, William. (1984). *Scientists, Society and the State: The Social Relations of Science Movement in Great Britain 1931–1947*. Columbus: Ohio State University Press.

McRae, Ron. (1984). *Mind Wars: The True Story of Secret Government Research into the Military Potential of Psychic Weapons*. St. Martin's Press.

McWhiney, Grady, and Jamieson, Perry. (1982). *Attack and Die: Civil War Military Tactics and the Southern Heritage*. Tuscaloosa: University of Alabama Press.

Mead, Walter. (1990). On the Road to Ruin: Winning the Cold War, Losing the Economic Peace, *Harper's*, March: 59–64.

Medawar, P. B. (1951). Is the Scientific Paper a Fraud?, *The Listener* 12, September: 377–378.

Mehan, Hugh, et al. (n.d.) The Politics of Representation in the Nuclear Arms Race, a proposal to the Institute for Global Conflict and Cooperation, University of California at San Diego.

Meistrich, Ira. (1991). The View Toward Armageddon, *MHQ: The Quarterly Journal of Military History* 3, Spring: 94–103.

Melman, Seymour. (1974). *The Permanent War Economy: American Capitalism in Decline*. Simon & Schuster.

Mendelssohn, Kurt. (1976). *The Secret of Western Domination: How Science Became the Key to Global Power, and What It Signifies for the Rest of the World*. Praeger.

Merchant, Carolyn. (1982). *The Death of Nature: Women, Ecology, and the Scientific Revolution*. Harper & Row.

Meyer, Rear Adm. Wayne. (1988). Interview, *Proceedings—US Naval Institute*, October: 102–106.

Military Space Staff. (1988a). Strategists See Space as Key to US Security, *Military Space* 5, February 1: 1.

———. (1988b). Computing Remains Toughest SDI Challenge, *Military Space* 5, May 9: 1.

———. (1988c). *Military Space* 5, August 6: 2.

———. (1988d). Supporting US Strategy for 3rd World Conflict, *Military Space* 5, August 29: 8.

———. (1988e). France Calls for Euro "Space Mastery," *Military Space* 5, October 24: 1.

———. (1989a). *Military Space* 6, June 19: 5.

———. (1989b). Stategist Calls for Space War Force, *Military Space* 6, August 28: 3.

———. (1989c). *Military Space* 6, October 23: 6.

———. (1990a). Brilliant Pebbles, *Military Space* 7, February 26: 1, 7–8.

———. (1990b). Global Changes Alter SDI Policy, *Military Space* 7, March 12: 3.

———. (1990c). Space Proliferation Worries DoD, *Military Space* 7, March 26: 3.

———. (1990d). *Military Space* 7, April 9: 6.

———. (1990e). *Military Space* 7, July 2: 3.

———. (1990f). *Military Space* 7, November 6: 2.

Military Technology Staff. (1987). Totally Automatic Weapons, *Military Technology* 5: 52.

Miller, Carolyn. (1990). The Rhetoric of Decision Science, or, Herbert A. Simon Says, in Herbert W. Simons, ed., *The Rhetorical Turn: Inventions and Persuasion in the Conduct of Inquiry*. Chicago: University of Chicago Press.

Miller, Marc. (1988). New Toys for Robocop Soldiers, *The Progressive*, July: 18–21.

Millis, Walter. (1956). *Arms and Men: A Study of American Military History*. Mentor Books.

Minsky, Marvin. (1982). Why People Think Computers Can't, *AI Magazine*, Fall: 32–33.

———. (1985). *The Society of Mind*. Touchstone.

———. (1989). The Intelligence Transplant, *Discover*, October: 52–58.

Moore, Jonathan, and Sciacchitano, Katherine. (1986). La Guerra de la Frontera: Imaginary Dangers—Real Militarization, *The National Reporter* 10, no. 2: 8–9.

Moore, Molly. (1987). Behind the Failure of the US Frigate, *San Francisco Chronicle*, May 25: 13–14.

Moorer, Adm. Thomas H., ret. (1987). Lessons Learned in the Air War Over Vietnam, *Proceedings—US Naval Institute*, August: 6.

Morgan Cmdr. John, Jr. (1988). DDG-51: Future Surface Force Prototype, *Proceedings—US Naval Institute*, December: 58–61.

Morgan, Robin. (1989). *The Demon Lover: On the Sexuality of Terrorism*. Norton.

Morgenthau, Hans. (1962). *Politics in the Twentieth Century*. Chicago: University of Chicago Press.

Morrison, David. (1990). Today's Threat to US Military: Prospect of Slashed Budgets Creates New Prototypes, *San Francisco Chronicle*, Briefing, March 4: 4.

Morse, Rob. (1991). Instant Quagmire: Just Add Sand, *San Francisco Examiner*, February 3: A3.

Morton, Janet. (1988). AI—Where It's Been, Where It's Going, *Defense Computing* 1, no. 2, March/April: 39–43.

Mueller, John. (1989). *Retreat from Doomsday: The Obsolescence of Major War*. Basic Books.

Mueller-Vollmer, Kurt. (1985). Language, Mind, and Artifact: An Outline of Hermeneutic Theory Since the Enlightenment, in K. Mueller-Vollmer, ed., *The Hermeneutic Reader*. Continuum.

Mulkay, Michael, Potter, Jonathan, and Yearley, Steven. (1983). Why an Analysis of Scientific Discourse Is Needed, in K. Knorr-Cetina and M. Mulkay, eds., *Science Observed: Perspectives on the Social Study of Science*. London: Sage, pp. 171–204.

Munro, Neil. (1995). A Look at the On-Line Frontier: The Pentagon's New Nightmare—an Electronic Pearl Harbor, *Washington Post*, July 16.

Mumford, Lewis. (1934). *Technics and Civilization*. Harcourt Brace Jovanovich.

———. (1970). *The Myth of the Machine*, 2 vols.: Vol. 1, *Technics and Human Development*; Vol. 2, *The Pentagon of Power*. Harcourt Brace Jovanovich.

Myles, Achebe. (1991). Operation Dope Storm: The Militarization of the War on Drugs, *City on a Hill*, March 14: 15.

Nadel, Alan. (1995). *Containment Culture*. Durham, N.C.: Duke University Press.

Nef, John. (1963). *War and Human Progress*. Norton.

Nelson, Greg, and Redell, David. (1986). The Star Wars Computer System, *Abacus* 3, no. 2, Winter: 8–22.

Nelson, Keith, and Olin, Spencer, Jr. (1979). *Why War? Ideology, Theory, and History*. Berkeley: University of California Press.

Newell, Lt. Col. Clayton. (1986). Operating in the 21st Century, *Military Review*, September: 4–10.

New York Newsday Staff. (1991). " 'Smart' Bombs Often Go 'Dumb,' " Experts Say, *San Francisco Chronicle*, February 11: A9.

Newsweek Staff. (1991a). Buzzwords, *Newsweek*, February 18: 12.

———. (1991b). Perspectives, *Newsweek*, February 18: 23.

Nideffer, Robert F. (1993). "Imag(in)ed Gulfs," CTheory Electronic Publication.

Nikutta, Randolph. (1987). Artificial Intelligence and the Automated Tactical Battlefield, in Din, 1987, pp. 100–130.

Nilsson, Nils J. (1995). Eye on the Prize, *AI Magazine* 16, no. 2, Summer: 9–17.

Noble, David. (1977). *America by Design*. Knopf.

———. (1986a). Command Performance: A Perspective on Military Enterprise and Technological Change, in Smith, M. R., 1985a, pp. 329–346.

———. (1986b). *Forces of Production*. Oxford: Oxford University Press.

Nolan, Jane E. and Wheelon, Albert D. (1990). Third World Ballistic Missiles, *Scientific American* 263, no. 2: 34–40.

Norris, Christopher. (1994). *Uncritical Theory: Postmodernism, Intellectuals, and the Gulf War*. London: Lawrence and Wishart.

North, David. (1986). US Using Disinformation Policy to Impede Technical Data Flow—Weapons Deception Plan, *Aviation Week & Space Technology*, March 17: 16.

Nyquist, Vice Adm. John. (1988). Interview, *Proceedings—US Naval Institute*, December: 75–79.

O'Connell, Charles, Jr. (1985). The Corps of Engineers and the Rise of Modern Management, 1827–1856, in Smith, M. R., 1985a, pp. 87–117.

O'Connell, Robert. (1989). *Of Arms and Men: A History of War, Weapons, and Aggression*. Oxford: Oxford University Press.

O'Connell, Teresa. (1990). From Scorched Earth to Scourged Earth: Drugs and Destabilization in Guatemala, unpublished ms.

Ornstein, Severo, Smith, Brian C., and Suchman, Lucy. (1984). Strategic Computing, *Bulletin of the Atomic Scientists*, December: 17–24.

Ornstein, Severo, and Suchman, Lucy. (1985). Reliability and Responsibility, *Abacus* 3, no. 1, Fall: 57–61, 68.

Owens, Adm. William A. (1995). The Emerging System of Systems, *Military Review*, May–June: 15–19.

Parker, Geoffrey. (1988). *The Military Revolution: Military Innovation and the Rise of the West, 1500–1800*. Cambridge, England: Cambridge University Press.

Parnas, David. (1985). Software Aspects of Strategic Defense Systems, *American Scientist*, September–October: 37–46.

———. (1987). Computers in Weapons: The Limits of Confidence, in D. Bellin and G. Chapman, eds., *Computers in Battle*. Harcourt Brace Jovanovich, pp. 202–232.

Parnas, David, and the National Test Bed Study Group. (1988). The SDI's National Test Bed: An Appraisal, CPSR, Inc., Report no. WS-100-5, May, Palo Alto, CA.

Partridge, Eric. (1966). *Origins: A Short Etymological Dictionary of Modern English.* Macmillan.

Patton, Phil. (1996). Robots with the Right Stuff, *Wired,* March: 148–151, 210–213.

Payne, R., and Hauty, G. (1954). The Effects of Experimentally Induced Altitudes on Task Efficiency, *Journal of Experimental Psychology* 47: 267–273.

Perkovich, George. (1987). Is Defense Reporting a Martial Art?, *Deadline,* July–August: 8.

Perlman, David. (1988). Major Advances in Nuclear Arms, *San Francisco Chronicle,* February 16: A18.

Perrin, Noel. (1979). *Giving Up the Gun: Japan's Reversion to the Sword, 1543–1879.* Berkeley, Cal.: Shambala Press.

Peters, Capt. Ralph. (1987). The Army of the Future, *Military Review,* September: 36–45.

Pollack, Andrew. (1989). Pentagon Sought Smart Truck but It Found Something Else, *New York Times,* May 30: 1A, C5.

Porter, Theodore M. (1995). *Trust in Numbers: The Pursuit of Objectivity in Science and the Public Life.* Princeton, N.J.: Princeton University Press.

Possony, Stefan, and Pournelle, Jerry. (1970). *The Strategy of Technology.* Cambridge, Mass.: University Press of Cambridge.

Preisel, Cmdr. J., Jr. (1988). High-tech Below the Main Deck, *Proceedings—US Naval Institute,* October: 121–124.

Pretty, Ronald, ed. (1987). *Jane's Weapon Systems 1986–87,* 17th ed. Jane's Publishing.

Pringle, Peter, and Arkin, W. (1983). *SIOP: Nuclear War from the Inside.* London: Sphere Books.

Pringle, Laurence. (1993). *Chemical and Biological Warfare: The Cruelest Weapons.* Springfield, N.J.: Enslow Publications.

Pullum, Geoffrey. (1987). Natural Language Interfaces and Strategic Computing, *AI and Society* 1, no. 1: 47–58.

Pylyshyn, Zenon, ed. (1987).*The Robot's Dilemma: The Frame Problem in Artificial Intelligence.* Norwood, N.J.: Ablex.

Quinn, Arthur. (1982). *Figures of Speech.* Salt Lake City: G. M. Smith.

Ractor. (1984). *The Policeman's Beard Is Half Constructed-Computer Prose and Poetry.* Warner Books.

Radine, Lawrence. (1977). *The Taming of the Troops: Social Control in the United States Army.* Westport, Conn.: Greenwood Press.

Rajchman, John. (1985). The Postmodern Museum, *Art in America,* October: 116–129.

Ramezoni, Ahmad. (1989). The Probability of Nuclear War in the Middle East, *Twanas,* Spring: 4–6.

Rangel, Charles. (1991). The Pentagon Pictures, *New Patriot* 4, no. 2, March–April: 12–13.

Rapaport, Anatol. (1962). Introduction, in Carl von Clausewitz, *On War.* Baltimore: Penguin.

Reed, Fred. (1987). The Star Wars Swindle: Hawking Nuclear Snake Oil, *Harper's,* May: 39–48.

Renique, Gerardo. (1990). "War Against Drugs" or "Dirty War," *The Mobilizer*, Summer: 10–11.

Restivo, Sal. (1988). Modern Science as a Social Problem, *Social Problems* 35, no. 3, June: 206–225.

Reston, James, Jr. (1984). *Sherman's March and Vietnam*. Macmillan.

Reuters Service. (1991). US, Allies Ready Markets for War: Emergency Plans Are Being Installed to Prevent Panic, *San Francisco Chronicle*, January 16: C1, C4.

Richardson, J. Jeffrey, ed. (1985). *Artificial Intelligence in Maintenance*. Park Ridge, N.J.: Noyes Publications.

Richey, Warren. (1986). Smart Computers to Aid FBI's Fight Against Crime, *Christian Science Monitor*, March 5: 3.

Robins, Kevin. (1991). The Mirror of Unreason, *Marxism Today*, March: 42, 44.

Rogers, William. (1969). *Think: A Biography of the Watsons and IBM*. Signet.

Rohatyn, Dennis. (1985). Butler and Turing: The Search for Superlife, in Bart Thurber et al., eds., *In Our Own Image: Readings in the Humanities and Technology*. Needham, Mass.: Ginn Press.

Rose, Frank. (1985). *Into the Heart of the Mind: An American Quest for Artificial Intelligence*. Vintage.

Rose, Hilary. (1983). Hand, Brain, and Heart: Toward a Feminist Epistemology for the Natural Sciences, *Signs*, Fall: 73–90.

Rose, Hilary, and Rose, Steven, eds. (1976). *The Radicalization of Science*. Macmillan.

Roszak, Theodore. (1986). *The Cult of Information*. Pantheon Books.

Roth, Bernard. (1987). Robots and the Military, *CPSR Newsletter* 5, no. 3, Fall: 1–6.

Royce, Knut, and Eisner, Peter. (1990). US Led Drug Raid in Colombia: Bush Denies It, but Evidence on Assault Piles Up, *San Francisco Chronicle*, May 4: A1, A24.

Salisbury, Harrison. (1967). *Behind the Lines—Hanoi*. Bantam.

San Francisco Chronicle Staff. (1991). Impartial Review of Gulf Victory, *San Francisco Chronicle*, May 8: A24.

San Jose Mercury News Staff. (1987). Army Seeks to Modernize Remains Identification, *San Jose Mercury News*, February 17: A12.

———. (1991). Gulf Quotes, *San Jose Mercury News*, February 28: A15.

Sanger, David. (1985). Campuses' Role in Arms Debated, *New York Times*, July 22: A1, A11.

Santa Rosa Press Democrat Staff. (1991). Football Is Obviously the Military's Sport of Choice, *Sonoma County Peace Press* 5, no. 4, May: A1.

Sarkesian, Sam, ed. (1980). *Combat Effectiveness: Cohesion, Stress, and the Volunteer Military*. London: Sage.

Satchell, Michael. (1988). The New US Military: Pride, Brains and Brawn, *San Francisco Chronicle*, Briefing, April 20: 2.

Scarry, Elaine. (1985). *The Body in Pain: The Making and Unmaking of the World*. Oxford: Oxford University Press.

Schank, Roger, and Abelson, Robert. (1977). *Scripts, Plans, Goals and Understanding*. New Haven, Conn.: LEA.

Scheer, Robert. (1982). *With Enough Shovels: Reagan, Bush & Nuclear War*. Random House.

Schell, Jonathan. (1982). *The Fate of the Earth*. Avon.

Schrage, Michael. (1986). Big Floyd, Software Special Agent, *San Francisco Sunday Chronicle-Examiner World Magazine*, August 24: 15.

Schwartz, Charles. (1988). Scientists: Villains and Victims in the Arms Race: An Appraisal and a Plan of Action, *Bulletin of Peace Proposals* 19, nos. 3–4: 399–409.

Scientific American Staff. (1987). Networking, *Scientific American* 256, January: 39.

Scriven, Michael. (1953). The Mechanical Concept of Mind, *Mind* 62, no. 246: 297–312.

Searle, John. (1980). Minds, Brains, and Programs, *The Behavioral and Brain Sciences* 3, September: 417–457.

Seidel, Gill. (1985). Political Discourse Analysis, in N. M. Van Dijk, ed., *Handbook of Discourse Analysis*. London: Academic Press.

Seiffert, Siegfried. (1987). Technical Interoperability of the Command, Control and Information System in the Central Region, *Military Technology* 11, no. 5: 93–95.

Shah, Vivek, and Buckner, Gary. (1988). Potential Defense Applications of Expert Systems, *IEEE AES Magazine*, February: 15–21.

Shapiro, Michael. (1981). *Language and Political Understanding—The Politics of Discursive Practices*. New Haven, Conn.: Yale University Press.

Shapley, Deborah. (1974). Weather Warfare: Pentagon Concedes 7-Year Vietnam Effort, *Science*, 7 June: 1059–1061.

Sheehan, Neil. (1988). *A Bright Shining Lie: John Paul Vann and America in Vietnam*. Random House.

Sherry, Michael. (1977). *Preparing for the Next War: American Plans for Postwar Defense 1941–45*. New Haven, Conn.: Yale University Press.

———. (1987). *The Rise of American Air Power: The Creation of Armageddon*. New Haven, Conn.: Yale University Press.

Siegel, Lenny. (1991). The Chemical Wastes of Warfare, *San Jose Mercury News*, February 6: A6, B11.

Simmons, Lewis. (1991). Anti-US Sentiment Brews in Muslim Asia, *San Jose Mercury News*, January 29: 14A.

Simon, Herbert A. (1980). Cognitive Science: The Newest Science of the Artificial, *Cognitive Science* 4, no. 2, April–June: 33–46.

Simon, Herbert A., and Newell, Allen. (1963). Computer Science as Empirical Inquiry: Symbols and Search, *Communications of the ACM*: 113–126.

Skelly, James. (1986). Power/Knowledge: The Problems of Peace Research and the Peace Movement, paper presented at the International Peace Research Association meeting, April, p. 10.

Skocpol, Theda. (1983). France, Russia, China: A Structural Analysis of Social Revolutions, in Jack Goldstone, ed., *Revolutions*, pp. 68–85.

Sloyan, Patrick. (1991). Biggest Battle Came After Truce: Damage Was Played Down by Military, *San Francisco Chronicle*, May 8: 1A, 14A.

Smith, Brian. (1985). The Limits of Correctness in Computers, in Charles Dunlop and Rob Kling, eds., *Computerization and Controversy: Value Conflicts and Social Choices*. Academic Press.

Smith, E., and Whitelaw, J. (1987). Artificial Intelligence Research in Australia—A Profile, *AI Magazine* 8, no. 2: 77–81.

Smith, Merritt Row, ed. (1985a). *Military Enterprise and Technological Change: Perspectives on the American Experience*. Cambridge, Mass.: MIT Press.

———. (1985b). Army Ordnance and the "American System" of Manufacturing, 1815–1861, in Smith, M. R., 1985a, pp. 39–86.

Smith, R. Harris. (1977). *OSS: The Secret History of America's First Central Intelligence Agency*. Delta Books.

Smith, R. Jeffrey. (1985). Weapons Labs Influence Test Ban Debate, *Science*, 12 September: 1067–1069.

Smoler, Fredric. (1989). The Secret of the Soldiers Who Didn't Shoot, *American Heritage* 40, no. 2, March: 36–45.

Snow, Lt. Col. Joel. (1987). AirLand Battle Doctrine Tenets in Opposition, *Military Review*, October: 63–67.

Sofia, Zoë. (1983). Towards a Sexo-Semiotics of Technology, Qualifying Essay for the History of Consciousness Board of Studies, University of California at Santa Cruz.

———. (1984). Exterminating Fetuses, *diacritics* 14, no. 2, Summer: 47–59.

Speer, Albert. (1970). *Inside the Third Reich: Memoirs*, Richard Winston and Clara Winston, trans. Macmillan.

Squires, Sally. (1988). The New Army Has Gone New-Age, *San Francisco Examiner-Chronicle, Punch*, May 1: 3.

Stallworthy, Jon. (1984). *The Oxford Book of War Poetry*. Oxford: Oxford University Press.

Steele, Jack E. (1996). Interview with Chris Hables Gray, in C. H. Gray, H. Figueroa-Sarrier, and S. Mentor, eds., *The Cyborg Handbook*. Routledge.

Stein, Kenneth. (1985). Humans Simulate Computer Logic in Expert System Project, *Aviation Week & Space Technology*, April 22: 73.

Stent, Gunter. (1986). Hermeneutics and the Analysis of Complex Biological Systems, *Proceedings of the American Philosophical Society* 130, no. 3: 336–342.

Sterling, Bruce. (1993). SimWar, *Wired*, May: 52–56.

Stewart, Jim. (1988). Warriors Without War, *This World*, April 24: 19–20.

Stewart, Jon. (1988). New Push for Usable Nuclear Weapons, *San Francisco Chronicle, Briefing*, February 17: 1.

Sugawara, Sandra. (1990). Tough Times for Trainers, *Washington Post, Washington Business*, September 10: 1, 30–1.

Sullivan, Gen. Gordon R. (1995). A Vision for the Future, *Military Review*, May–June: 4–14.

Sun, Tzu. (1962) *The Art of War*, S. B. Griffith, trans. Oxford: Oxford University Press.

Swett, Charles. (1995). *Strategic Assessment: The Internet*. Washington, D.C.: Office of the Assistant Secretary of Defense for Special Operations and Low Intensity Conflict.

Szafranski, Col. Richard. (1995). A Theory of Information War: Preparing for 2020, *Airpower Journal* IX, no. 1 (Spring): 18-27.

Tax, Mendit. (1993). Five Women Who Won't Be Silenced, *The Nation*, May 10: 626.

Theweleit, Klaus. (1989). *Male Fantasies*, Vol. 1: *Women Floods, Bodies History*; Vol. 2: *Male Bodies: Psychoanalyzing the White Terror*. Minneapolis: University of Minnesota Press.

Thomas, Evan, and Barry, John. (1991). War's New Science, *Newsweek*, February 18: 39.

Thompson, Clark. (1984). Federal Support of Academic Research in Computer Science, Computer Science Division report, University of California at Berkeley.

Thompson, E. P., and Smith, Dan, eds. (1981). *Protest and Survive*. Monthly Review Press.

Thompson, Mark. (1988). How Army's Plans for Drone Turned into a $2 Billion Drain, *San Jose Mercury News*, January 11: 1.

———. (1995). If War Comes Home, *Time*, August 21: 44–46.

Thucydides. (1982). *The Peloponnesian War*, R. Crawley, trans. Modern Library.

Timmerman, Col. Frederick, Jr. (1987). Future Warriors, *Military Review*, September: 32–56.

Tirman, John, ed. (1984). *The Militarization of High Technology*. Ballinger.

Toffler, Alvin, and Toffler, Heidi. (1993). *War and Anti-War*. Warner Books.

Toulmin, Stephen. (1982). The Construal of Reality: Criticism in Modern and Postmodern Science, *Critical Inquiry* 9, September: 93–111.

Trewhitt, Henry. (1971). *McNamara*. Harper & Row.

Truver, Scott. (1988). Whither the Revolution at Sea? *Proceedings—US Naval Institute*, December: 68–74.

Trux, Jon. (1991). Desert Storm: A Space-Age War, *New Scientist* 27, July: 30–34.

Tumey, Capt. David. (1990). Mind Over Matter, *Discover*, August: 16.

Turing, Alan. (1950). Computing Machinery and Intelligence, *Mind* 59, no. 236: 433–460.

Tyler, Patrick. (1992). Pentagon Paper Says Soviets Are Still a Threat in Third World, *San Francisco Chronicle*, February 13: A13.

United Press Service. (1989). US Permission to "Shoot First" Biggest Change, Says Admiral, *Seattle Times*, September 2: A4.

———. (1991). Most of Boeing Contract in Saudi Project Voided, *San Francisco Chronicle*, January 11: A19.

Urioste, Lt. Cmdr. Marcus. (1988). Where Is the SSN Going?, *Proceedings—US Naval Institute*, October: 109–112.

Vagts, Alfred. (1937, 1959) *A History of Militarism*. Meridian.

Van Creveld, Martin. (1989). *Technology and War: From 2000 B.C. to the Present.* Free Press.

———. (1985). *Command in War*. Cambridge, Mass.: Harvard University Press.

———. (1991). *The Transformation of War*. Free Press.

Van Dijk, Tuen, ed. (1985). *Handbook of Discourse Analysis*. London: Academic Press.

Vickers, Earl. (1984). War and Games, *Creative Computing*, September: 146–153.

Virilio, Paul. (1986). *Speed and Politics: An Essay on Dromology*, M. Polizzotti, trans. Semiotext(e).

———. (1990). *War and Cinema: The Logistics of Perception*, P. Camiller, trans. London: Verso.

Virilio, Paul, and Lotringer, Sylvere. (1983). *Pure War*, M. Polizotti, trans. Semiotext(e).

Vitale, Bruno. (1985). Scientists as Military Hustlers, in Radical Science Collective, eds., *Issues in Radical Science*. London: Free Association Press.

Waller, Douglas, et al. (1990). Inside the Invasion, *Newsweek*, June 25: 28–31.

———. (1995). Onward Cybersoldiers, *Time*, August 21: 38–44.

Walt, Stephen. (1987). The Search for a Science of Strategy, *International Security* 12, no. 1, Summer: 40–65.

Warren, Jim. (1996). Surveillance-on-Demand, *Wired*, February: 72–73.

Watson, Russel, et al. (1995). When Words Are the Best Weapon, *Newsweek*, February 27: 36–40.

Webb, Gary. (1991). US Scuttles Suits Alleging Naval Gun Doesn't Work, *San Jose Mercury News*, February 21: 1A, 10A.

Wehling, Jason. (1995). Netwars and Activists, Power on the Internet, *CTheory* (www.ctheory.com), Electronic.

Weinberger, Caspar. (1988). Basic Trends Shaping the Security Environment and the Battlefield of the Future, in S. Pfaltzgraff, Jr., et al., eds., *Emerging Doctrines and Technologies: Implications for Global and Regional Political–Military Balances*. Lexington, Mass.: Lexington Books, pp. 3–10.

Weiner, Tim. (1987). Pentagon Plans for WWIV, *San Jose Mercury News*, February 16: A1, A10.

———. (1988). "Black Budget" series, *San Jose Mercury News*, February 13–16: all on page A1.

———. (1990). *Blank Check: The Pentagon's Black Budget*. Warner Books.

———. (1991). Pentagon Budget Hides $34 Billion from the Public, *San Jose Mercury News*, February 6: A1, A7.

———. (1992). Expert Says "Smart Bombs" Led to 70,000 Civilian Deaths, *San Jose Mercury News*, January 9: A4.

Weisbin, C. (1987). Intelligent Machine Research at CESAR, *The AI Magazine* 8, no. 1: 62–74.

Weizenbaum, Joseph. (1976). *Computer Power and Human Reason*. San Francisco: Freeman.

Westmoreland, William. (1969). Address to the Association of the U.S. Army, October 14. Reprinted in Paul Dickson, *The Electronic Battlefield*. Athenum, 1975, pp. 256–264.

White, Hayden. (1973). *Metahistory: The Historical Imagination in Nineteenth Century Europe*. Baltimore: Johns Hopkins University Press.

———. (1978). *Tropics of Discourse: Essays in Cultural Criticism*. Baltimore: Johns Hopkins University Press.

———. (1986). *The Content of the Form: Narrative Discourse and Historical Representation*. Baltimore: Johns Hopkins University Press.

Whitefield, Stephen J. (1990). *The Culture of the Cold War*. Baltimore: Johns Hopkins University Press.

Wiener, Norbert. (1954). *The Human Use of Human Beings*. Houghton Mifflin.

Wilford, John. (1986). It's All in His Head, *San Francisco Chronicle-Examiner, Punch*, July 13: 5.

Wilson, Andrew. (1968). *The Bomb and the Computer: Wargaming from Ancient Chinese Mapboard to Atomic Computer*. Delacorte.

Wilson, J. R. (1988). Simulating Helicopter Combat Conditions, *Interavia*, November: 1127–1129.

Winkler, Karen J. (1995). "Timely Insight," *The Chronicle of Higher Education*, September 29: A12–13.

Winner, Langdon. (1986). *The Whale and the Reactor: A Search for Limits in the Age of High Technology*. Chicago: University of Chicago Press.

Winograd, Terry. (1987). Strategic Computing Research and the Universities, Silicon Valley Research Group Working Paper no. 7, March.

Winograd, Terry, and Flores, Fernando. (1986). *Understanding Computers and Cognition*. Norwood, N.J.: Ablex.

Wolf, Michael. (1969). Commentary, in Monte Wright and Lawrence Paszek, eds., *Science, Technology, and Warfare: The Proceedings of the Third Military History Symposium*. Washington, D.C.: U.S. Government Printing Office.

Wolfe, Alexander. (1985). Strategic C^3—More Is Less, *Electronics Week*, April 1: 18–21.

Wood, David. (1994). New War Order Erodes Power of U.S. Military, *The Oregonian*, A3.

Wright, Susan, ed. (1993). *Preventing a Biological Arms Race*. MIT Press.

Wright, Quincy. (1964). *A Study of War*, abridged ed. Chicago: Chicago University Press.

Yudken, Joel, and Simon, Barbara. (1989). A Field In Transition: Current Trends and Issues in Academic Computer Science, Final Report of the Project on Funding Policy in Computer Science, for the Science Policy Committee Special Interest Group on Automata and Computability Theory, Association for Computing Machinery. Washington, D.C.

Zakaria, Fareed. (1987). The Colonel's Coup, *This World*, August 2: 19.

Zakheim, Dov. (1988). Superpower Military Power Projection: New Technologies and Third World Conflict, in S. Pfaltzgraff, Jr., et al., eds., *Emerging Doctrines and Technologies: Implications for Global and Regional Political–Military Balances*. Lexington, Mass.: Lexington Books, pp. 235–250.

Zilsel, Edgar. (1942). The Sociological Roots of Science, *American Journal of Sociology* 47.

Ziman, John. (1976). *The Force of Knowledge: The Scientific Dimension of Society*. Cambridge, England: Cambridge University Press.

3. Government Documents

Congress of the United States. (1970). Testimony of General William Westmoreland, *Congressional Record*, S11104, July 13.

———. (1975). *Biomedical and Behavioral Research*, Joint Congressional Hearings, 94th Congress, First Session, September 10 to November 7. Washington, D.C.: U.S. Government Printing Office.

———. (1985). *Hearings on H.R. 1872*, Director of DARPA, Part 4, March–April. Washington, D.C.: U.S. Government Printing Office.

———. (1988). *Hearings on the Sale of the Aegis Weapon System to Japan* by the Seapower and Strategic and Critical Materials Subcommittee of the Committee on Armed Services, House of Representatives, 100th Congress, First and

Second Sessions, December 1, 1987, February 4, 11, and March 3, 1988, Washington, D.C.: U.S. Government Printing Office.

———. (1989). Investigation into the Downing of an Iranian Airliner by the USS *Vincennes*, Committee on Armed Services, U.S. Senate, September 8, 1988. Washington, D.C.: U.S. Government Printing Office.

DARPA. (1983). *The Strategic Computing Program—New Generation Computing Technology: A Strategic Plan for its Development and Application to Critical Problems in Defense*. Defense Advanced Research Projects Agency, U.S. Dept. of Defense.

———. (1986). *Strategic Computing Program Second Annual Report*. Defense Advanced Research Projects Agency, U.S. Dept. of Defense.

———. (1987). *Strategic Computing Program Third Annual Report*. Defense Advanced Research Projects Agency, U.S. Dept. of Defense.

Eastport Study Group. (1985). A Report to the Director of the Strategic Defense Initiative Organization. Washington, D.C.: U.S. Army.

Presidential Commission on Integrated Long-Term Strategy. (1988). *Discriminate Deterrence*, Washington, D.C.: U.S. Government Printing Office.

Presidential Commission on Integrated Long-Term Strategy, Future Security Environment Working Group. (1988). *The Future Security Environment: Report to the Commission on Integrated Long-Term Strategy*, Washington, D.C.: U.S. Government Printing Office.

U.S. Air Force. (1985). Futurists II Brought New Dimensions in Aerospace, News Release, Aeronautical Systems Division, Office of Public Affairs, Wright–Patterson AFB.

———. (1986). Forecast II Executive Summary [and project list].

———. (1987). Gen. Lawrence Skantze's speech, "AF Science & Technology—The Legacy of Forecast II," given at Aerospace '87, the AIAA convention, Crystal City, Va., April 29.

U.S. Army. (1982). *AirLand Battle 2000*, August. U.S. Dept. of Defense.

———. (1984). DARCOM Long Range Planning Team Report: Trends and Their Implications for DARCOM During the Next 2 Decades, *Army RD&A Magazine*, May–June: 20.

———. (1986). *Joint Low-Intensity Conflict Project Final Report*, Vol. I: *Analytical Review of Low-Intensity Conflict*, August 1. U.S. Dept. of Defense.

———. (n.d.). Handout signed by the Commandant, U.S. Army Quartermaster School, "Robotics/Artificial Intelligence Fact Sheet."

U.S. Navy. (1988). *Navy 21*, Naval Studies Board.

Index